Back to Basics

Setting up your Workshop

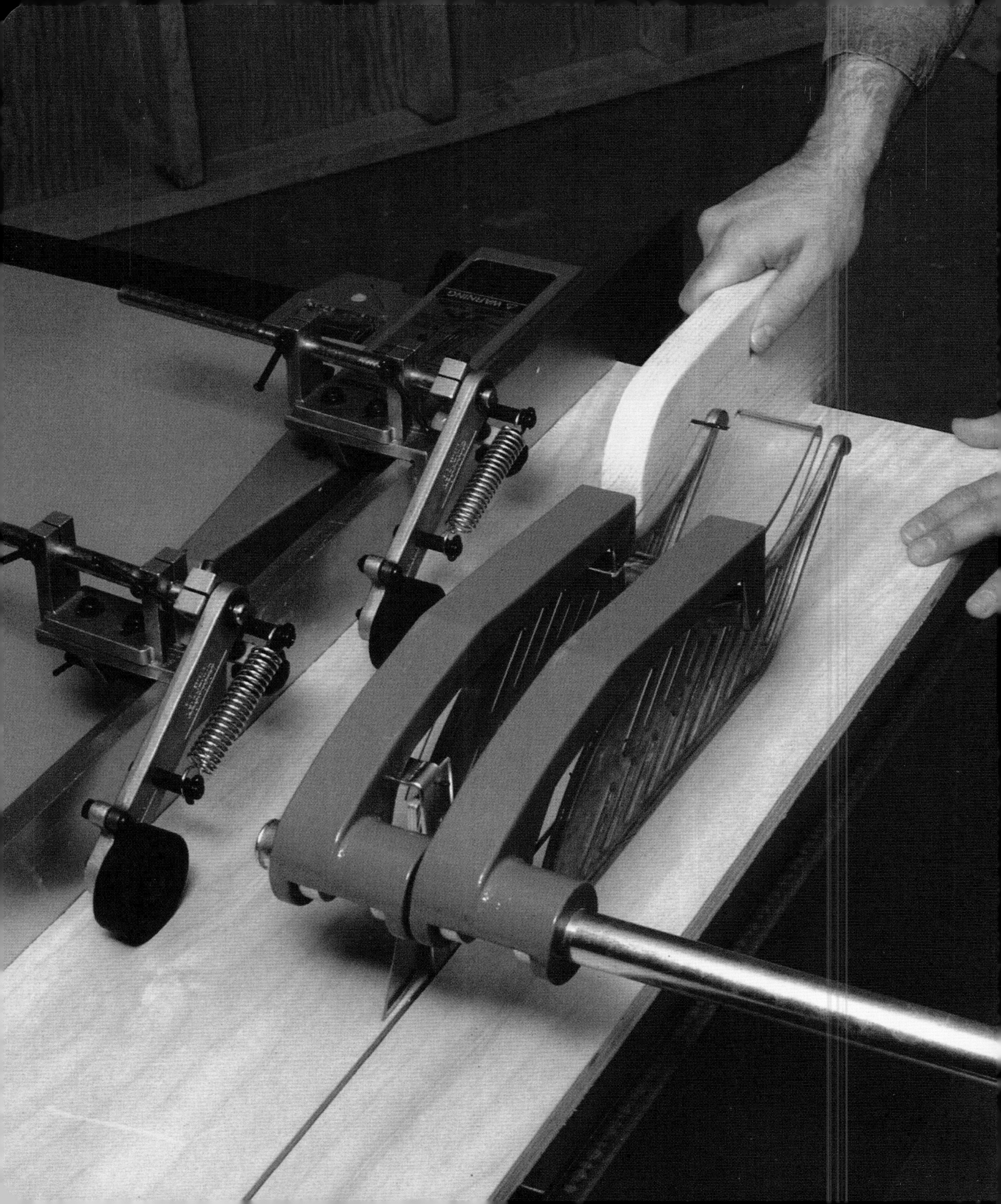
WARNING

Back to **Basics**

Setting up your Workshop

Straight Talk for Today's **Woodworker**

Published and distributed in North America by Fox Chapel Publishing Company, Inc.

Setting Up Your Workshop is an original work, first published in 2010.

Portions of text and art previously published by and reproduced under license with Direct Holdings Americas Inc.

ISBN 978-1-56523-463-5

Library of Congress Cataloging-in-Publication Data

Setting up your workshop.
p. cm. -- (Back to basics)

Includes index.

ISBN: 978-1-56523-463-5

1. Workshops--Equipment and supplies. 2. Woodworking tools. I. Fox Chapel Publishing.

TT152.S47 2010
684'.08--dc22

2010003679

To learn more about the other great books from Fox Chapel Publishing, or to find a retailer near you, call toll-free 800-457-9112 or visit us at *www.FoxChapelPublishing.com*.

Note to Authors: We are always looking for talented authors to write new books in our area of woodworking, design, and related crafts. Please send a brief letter describing your idea to Acquisition Editor, 1970 Broad Street, East Petersburg, PA 17520.

Printed in China
First printing: May 2010

Contents

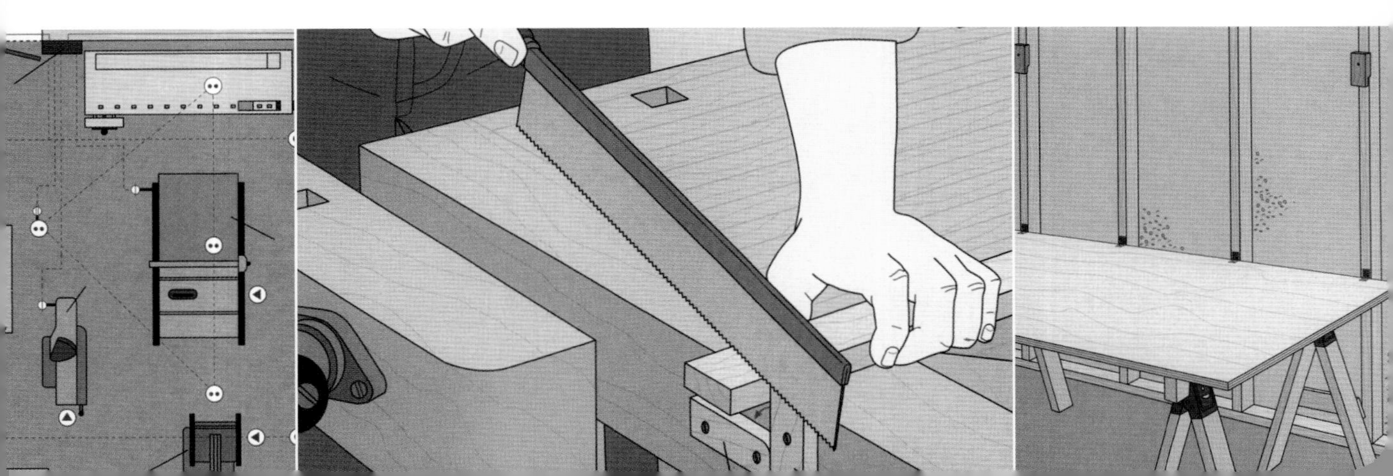

What You Can Learn

Shop Layout, p. 12

Few homes have space specifically designed as a workshop area. As a result, setting up a home shop demands creativity and flexibility.

Workbench, p. 30

With each refinement the workbench has assumed an increasingly indispensable role in the workshop.

Shop Accessories, p. 54

Look beneath the surface of an efficient, well-equipped shop, and you will find several invisible auxiliaries: accessories designed to make the work safer and the shop more comfortable in which to work.

Storage, p. 72

Adequate workshop storage should accomplish two goals: tools and materials should be kept within easy reach of each operation, and the storage devices should encroach as little as possible on work space.

Work Surfaces, p. 92

It is a truism that no workshop is ever large enough; it is equally true that no woodworker ever has enough tables, benches, sawhorses, stands, or props to support work in progress.

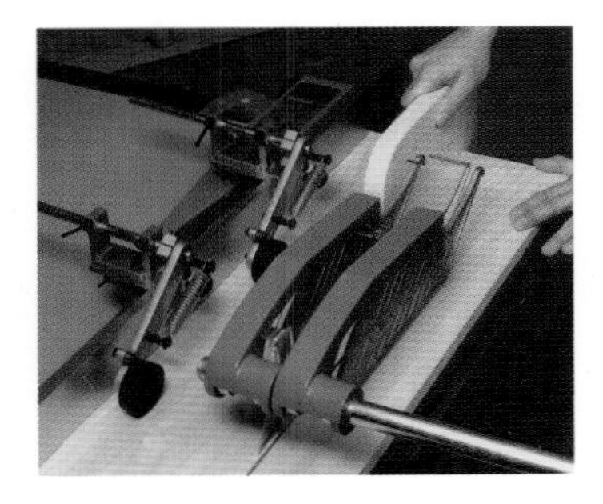

Safety, p. 130

The risk of mishap in the workshop can be reduced by a few simple precautions.

Dream Workshop

When I started woodworking in England, I was playing rock music and needed a hobby to help me unwind from the rigors of the road. My shop was just big enough for a workbench, a radial arm saw, and not much else. I remember constantly bumping things into the low ceiling.

When my wife and I moved back to America I was offered a job in a local wood-shop run by a friend of mine. It was there that I began to acquire a feel for how a shop should be laid out. After a few years with my friend, I left and set up my own shop in our three-car garage. I soon learned what worked and what didn't. When it came time to plan a shop from scratch, I sought out my local woodworking organization, which turned out to be a fountain of information. I studied lots of shops in my area and asked hundreds of questions about what people liked and what they wanted to change in their own shops.

Peter Axtell builds fine furniture at his shop in Sonoma County, California.

I saw one place in particular that seemed the best for my needs and settled on that as my model. My budget allowed me 1,500 square feet, so that was one limit set. Light—both natural and artificial—was a major consideration. I placed my building and planned the windows to take maximum advantage of the abundant sunlight in northern California. I was also determined to have a wood floor. I quickly found out that a hardwood floor would be too expensive so I used 1⅛-inch tongue-and-groove plywood and epoxy paint, which have held up very well.

I figured out the floor plan on graph paper and cut out scale drawings of all my machines as well as areas for plywood storage, office space, and a spray booth. It is important to allow enough space around your machines, so I spent considerable time movings things around and testing different scenarios.

The crawl space under my shop has extra clearance because I chose to run my dust collection pipe under the floor, which has allowed me to keep the whole ceiling space clear and airy. One of the best investments I made was in super-insulating the whole shop—floors, walls, and ceiling, which has made it easier to keep the place warm in winter and cool in summer.

I put a lot of thought and research into my shop and there isn't much I would change except for one thing: I wish I had built it bigger. But that is a common complaint. It seems that you can never have too much space.

- Peter Axtell

Shop Storage

Martha Collins designs and makes fine jewelry and furniture in her workshop near Sequim, Washington. Her husband, luthier Richard Schneider, works in an adjoining shop.

I make jewelry from exotic wood and dyed veneer. Some pieces have as many as 800 bits of wood in them, combining the colors and textures of various rare woods with brightly hued veneers. Being able to find some offbeat screw or fastener when I need it, or knowing where to retrieve that wonderful small chunk of rosewood that I've been saving for 10 years isn't a luxury; it's a necessity. Through the years, I have learned that the strength of a workshop depends on proper organization and storage.

I have four distinct "storage areas" in my shop. The infeed and outfeed tables of my radial arm saw hold scrap wood and less frequently used tools. My jewelry storage bench holds all the machined and milled exotic woods and dyed veneers. The 4-by-8 outfeed table on the table saw houses a variety of items: work in progress, exotic lumber, furniture pads, and leftovers from the jewelry-making process.

The most important storage area is in the main workbench and tool chest area. This is the heart of my shop. I keep everything from screws and screwdrivers to planes and hinges in cabinets close to the bench. All of the hand and power tools that I use regularly are kept in my main chest, which is featured in the photograph.

The chest is 6 feet high and 4 feet wide. In the upper section, I store everything from hammers and one set of chisels in the left-hand door to files, screwdrivers, and planes in the right-hand section. The five drawers in the center are filled with wrenches and bits of all kinds—spurs, twists, and Forstners. I assembled the drawers with dovetail joints, a satisfying effort that only adds to the pleasure of putting things away.

The left-hand door in the bottom section of the chest holds all my measuring and marking tools; the right-hand door houses a set of pliers along with my handsaws. The cubbyholes are home to my portable power tools—circular saws, saber saws, router, sander, cordless drills, pneumatic tools, and so on.

The tools in my chest have changed over the years. Fifteen years ago I had only one cordless drill; now I have three. But my chest has been able to adapt and accommodate all the new tools—each with its own specific place.

- Martha Collins

The Workbench

A workshop can be anywhere you can fit a solid surface. A retired carver friend built a superb workshop in the linen closet of his apartment. He only had to open the closet door, pull out a stool, and go to work. Everything he needed was fitted into a space of less than 10 square feet.

I built the small cherry bench in the photograph to fit an awkward alcove in my office that measures only 23 by 37 inches. For years I had been using my desk as a makeshift workbench and I was frustrated by both the lack of any decent clamping system and enough clear work surface. The desk is often as cluttered as the bookcase in the background.

With the workbench in place, I can now clamp wood for testing saws, chisels, bits, and so on, without knocking a coffee cup to the floor or spilling papers everywhere. The bench is also just the right height for using an inspection microscope, an invaluable tool for analyzing failures and successes in the world of sharp edges.

The bench occupies an otherwise unusable space next to a doorway. Since the floor space next to it can be used only for foot traffic, the bench only adds to the usability of my office; it does not detract anything. Incidentally, the bench was pulled out of the alcove for this photo.

More important than its utility, my bench adds a wonderfully relaxing and humanizing element. Like many people, I tire quickly of administrative detail. With a workbench handy, I can get up from my desk, wander over to the bench and tinker with tools for a while. It is like a mini-vacation in the middle of the day.

The humanizing part comes from surrounding yourself with things you like. I like everything about woodworking. My office is filled with old tools as well as books about their history and use. To add a workbench to the general clutter is just another layer to the cocoon. The world looks much better when viewed from an office with a workbench in it.

- Leonard Lee

Leonard Lee, recently retired president of Veritas Tools and Lee Valley Tools in Ottawa, Canada, manufacturers and retailers of fine woodworking hand tools.

Shop Layout

As they gain experience and accumulate tools, most woodworkers pine for their own special place to practice their skills. In their fantasies, the workshop is an airy space equipped with a substantial workbench and an array of stationary machines and portable tools. The reality for many woodworkers, however, is much more modest. The typical shop never seems to have enough light, power, or elbow room.

Few homes have space specifically designed as a workshop area. As a result, setting up a home shop demands creativity and flexibility; the task often involves converting an area originally intended for some other purpose. With careful planning and forethought, however, a location that might appear unsuitable can be turned into an efficient, comfortable place to work.

Although size is often the first consideration, several other concerns may be more important. For example, situating a shop in a spare room on the main floor of a home may provide a large working area, but noise and dust from tools would probably inconvenience other members of the family. To suit their own needs without intruding too much on the people they live with, woodworkers commonly locate home shops in the basement or a garage. Each has its pros and cons. A basement is apt to be damp and may need to have its wiring and heating upgraded; access can be hampered by narrow doors, tight stairways, and low ceilings; and ventilation may be inadequate for finishing tasks. A garage, on the other hand, is apt to be cold; it may require wiring and heating. The woodworker may end up jostling for space with a car or two.

Still, with a bit of planning and the proper layout, even these locations can be turned to your advantage: A basement can be heated and powered more easily than a garage. On the other hand, a garage has a larger door through which to move lumber and sheet materials like plywood, its air is less humid, and the din of power tools and fumes of finishing can be isolated from living spaces.

Even in spacious shops, tools occasionally need to be moved around; in small shops, reassigning floor space may be a part of every project. A wheeled base can make a 10-inch table saw, like the one pictured here, easy to reposition.

Every hour spent planning shop layout pays dividends later on. To determine the best way to arrange the tools planned for the shop, a woodworker places overhead-view silhouettes of the tools on a scale drawing of the space.

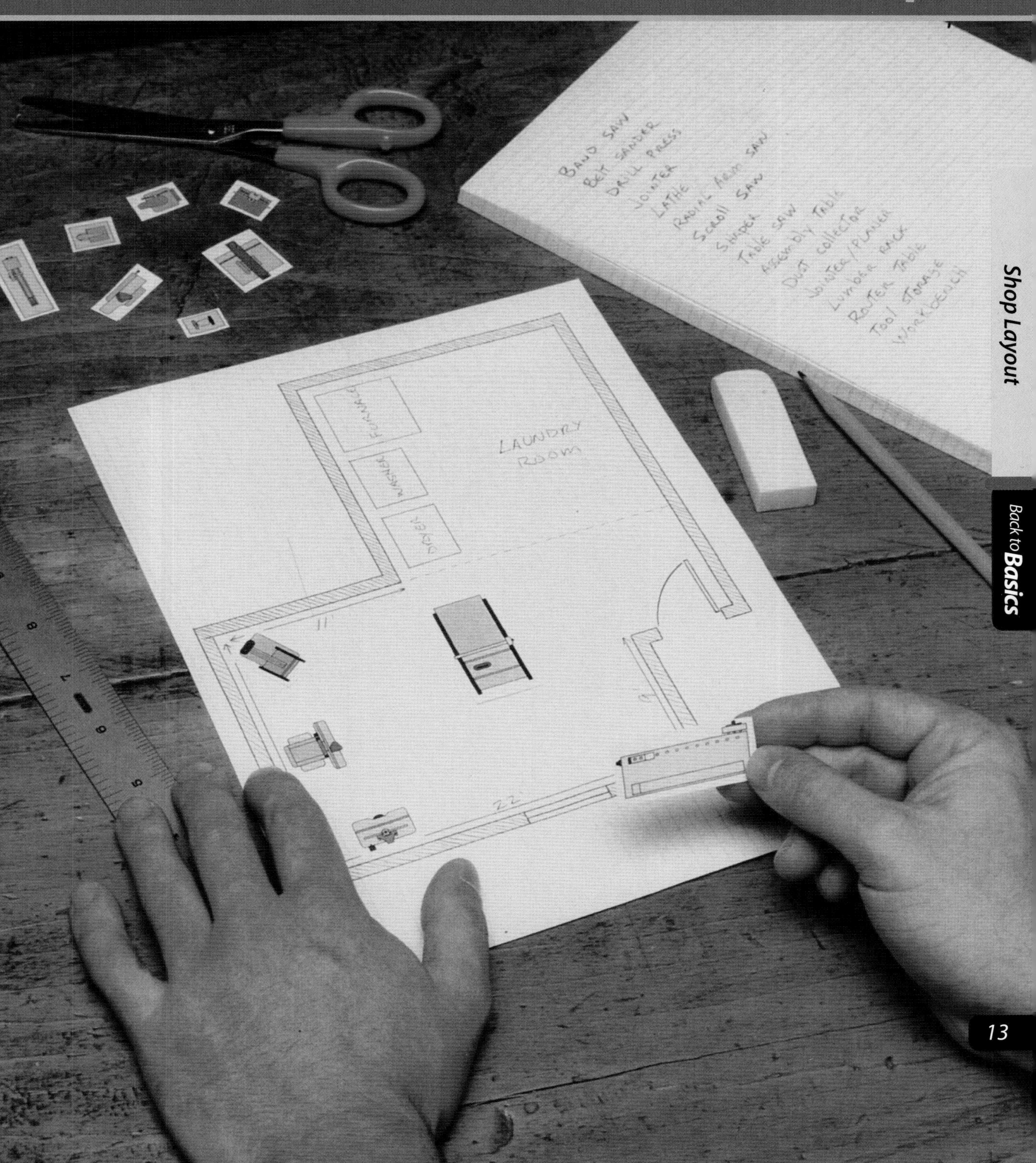
BAND SAW
BELT SANDER
DRILL PRESS
JOINTER
LATHE
RADIAL ARM SAW
SCROLL SAW
SHAPER
TABLE SAW
ASSEMBLY TABLE
DUST COLLECTOR
JOINTER/PLANER
LUMBER RACK
ROUTER TABLE
TOOL STORAGE
WORKBENCH
FURNACE
WASHER
DRYER
LAUNDRY ROOM

Workshop Planning

It is far easier to shuffle paper cutouts of your tools on a template than it is to drag a table saw halfway across the shop. Time spent planning the layout of your shop will be more than amply rewarded in reduced frustration and increased efficiency when you go to work.

Designing a shop involves juggling many interdependent variables, from local humidity and the type of work you do to the height of the ceiling and the cost of wiring. To help sort them out, ask yourself a set of questions, like those in the checklist on page 15, to help determine the kind of shop most suitable for your needs and remind you of factors that may affect its design. Remember, too, a basic principle for any shop, illustrated below, that the lumber should take a relatively straight path as it is processed—almost as though the shop were an assembly line.

Refer to the illustrated inventory of stationary machines and tables starting on page 16 as a guide to space and lighting requirements. The best way to design the layout is to experiment with arranging photocopies of scale drawings of the tools *(page 19)* on a sheet of graph paper. Remember that a tool should be positioned so that

Tool Placement and Work Flow

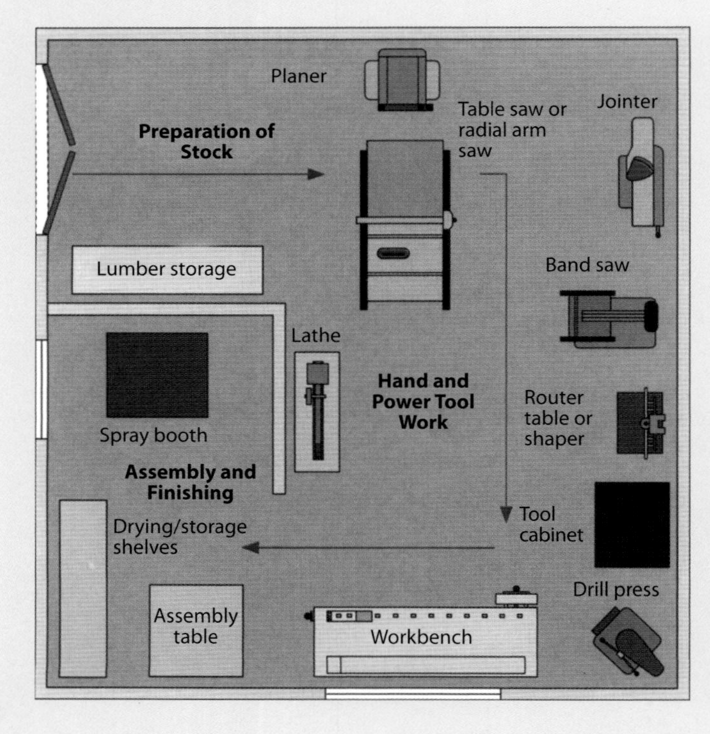

Designing a shop around the woodworking process
For maximum efficiency, lay out the tools in your shop so that the lumber follows a fairly direct route from rough stock to finished pieces. The diagram at left illustrates a logical work flow for a medium-size workshop. At the upper left-hand corner is the entrance where lumber is stored on racks. To the right is the stock preparation area, devoted to the table saw (or radial arm saw), jointer, and planer; at this station, lumber is cut to rough length and surfaced. The heart of the next work area, near the bottom right-hand corner of the drawing, is the workbench. Radiating outward from the bench are the shop's other stationary tools—in this case, a drill press, lathe, router table (or shaper), and band saw. A tool cabinet is nearby. Moving clockwise, the final work area is set aside for assembly and finishing. This station features a table for gluing up pieces and shelves for drying and storing. The spray booth is close by, but isolated from the shop by walls on three sides.

an access door is visible from it. In addition, a workpiece kicked back from the tool should not be able to strike someone working at another station.

Consider dedicating spaces for specific woodworking tasks. A finishing area or spray booth requires priority in planning because of light, temperature, and ventilation needs.

Depending on the extent of your shop and local zoning and building codes, you may need to obtain permits; consult your local building inspection office.

A Shop Layout Checklist

Location

- Which available areas in and around your home are appropriate for a shop?
- How easy is the access to these areas?
- Is the electric wiring adequate for powering your tools and lighting?
- How well are the areas heated, insulated, and ventilated?
- Will shop noise disturb other areas?
- If the location is a basement, will the shop be sharing space with a furnace room or laundry room?
- If the location is an outbuilding or garage, how much space is taken up by cars, bicycles, lawn mowers, and so on?
- Does the building or garage have any heating, electricity, or plumbing?
- How secure is the building or garage from theft?

Type of Work

- What type of woodworking projects will you be doing?
- What size are the materials you will need to move in and out of the shop?
- How much space will be devoted to storing lumber and work-in-progress?
- What stationary machines, portable power tools, and hand tools will you need?
- Are there enough electrical circuits to supply your power needs?
- How many lighting fixtures does your work require?
- How many workbenches, assembly tables, and accessories like tool cabinets, scrap bins, and sawhorses will you need?
- Will local seasonal temperatures and humidity affect your work?
- Will you be doing a lot of finishing work?

Work Habits

- What room temperature will you need to work comfortably?
- What type of light do you prefer for working?
- Will you be working during daylight hours, or will you be using the shop at night?
- Which tools do you expect to use most often?
- Will you be working alone in the shop, or will it be used by another worker?
- Would that person have easy access to the shop?
- Will you need to lock the shop or keep it off-limits to children or pets?
- How many hours per day do you expect to spend in the shop?
- Is the flooring made of a material that is comfortable to stand on for long periods of time?

Planning for Stationary Tools

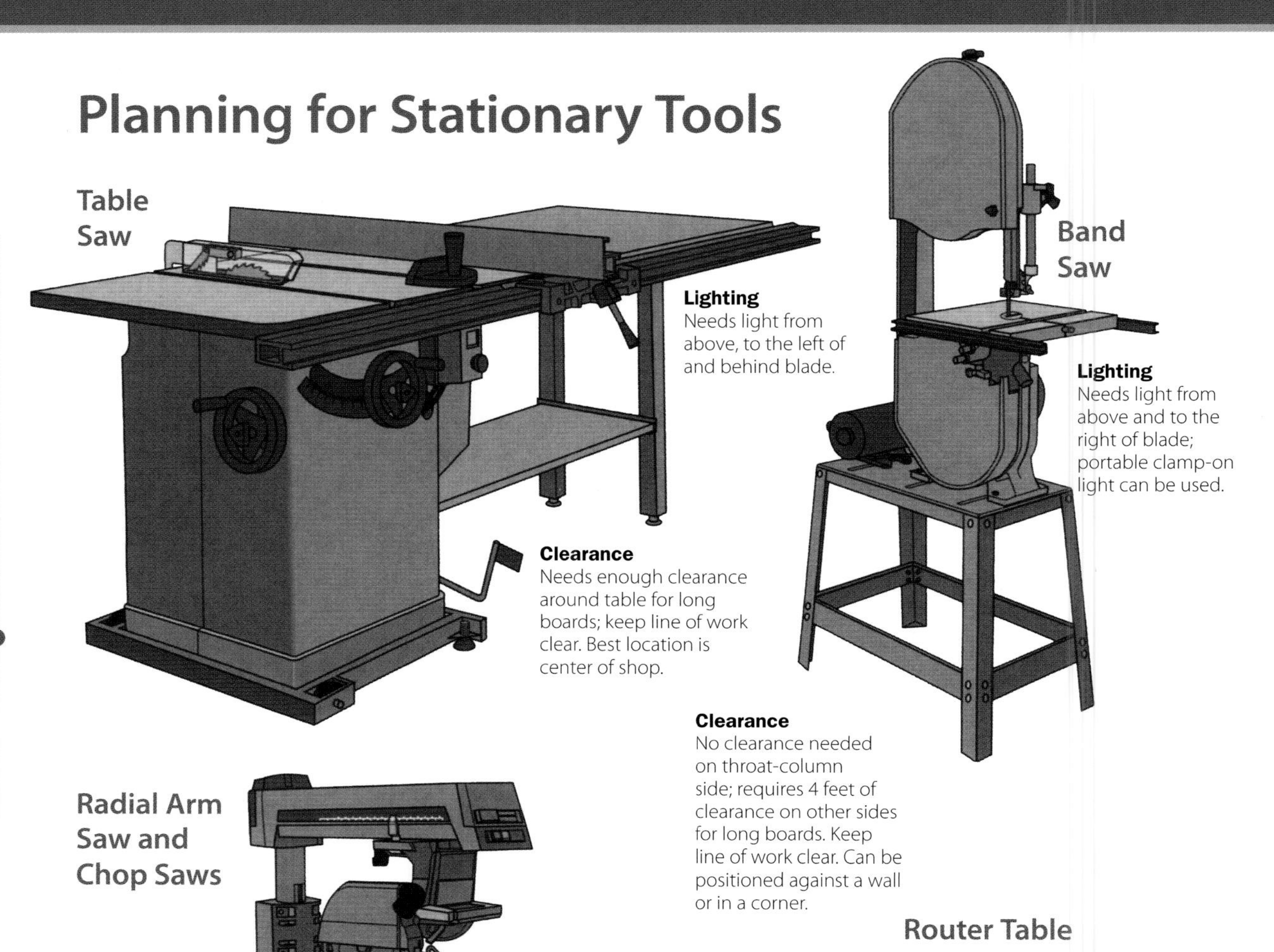

Table Saw

Lighting
Needs light from above, to the left of and behind blade.

Clearance
Needs enough clearance around table for long boards; keep line of work clear. Best location is center of shop.

Band Saw

Lighting
Needs light from above and to the right of blade; portable clamp-on light can be used.

Clearance
No clearance needed on throat-column side; requires 4 feet of clearance on other sides for long boards. Keep line of work clear. Can be positioned against a wall or in a corner.

Radial Arm Saw and Chop Saws

Lighting
Needs light from sides, front and above.

Clearance
No clearance needed behind tool; good location is against wall. Allow about 12 feet of clearance on either side for long boards.

Router Table

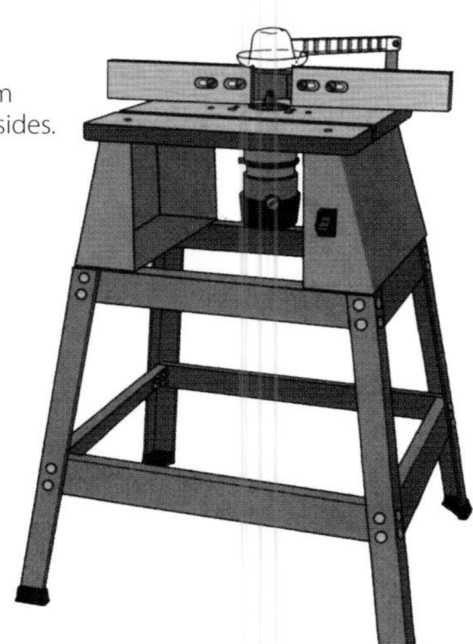

Lighting
Needs light from front and both sides.

Clearance
Relatively portable. Allow at least 6 feet of clearance in front and to the sides when in operation; needs no clearance behind table. Good location is against wall; keep line of work clear so that a kicked back workpiece would not strike another worker.

Shaper

Lighting
Needs light from front and sides.

Clearance
Allow at least 6 feet of clearance in front and to the sides; no clearance needed behind tool Good location is against wall; keep line of work clear so that a workpiece that is kicked back will not strike another worker.

Drill Press

Lighting
Light must focus directly on bit from overhead; portable clamp-on light can be used.

Clearance
No clearance needed behind tool; good location is against wall. Allow 3 feet of clearance on either side, and enough clearance in front for wide work.

Jointer/Planer

Lighting
Needs light from above and front of tables.

Clearance
Allow 3 to 4 feet of clearance on sides, more for long boards; keep line of work clear. Good location is near lumber storage rack for easy surfacing of stock.

Jointer

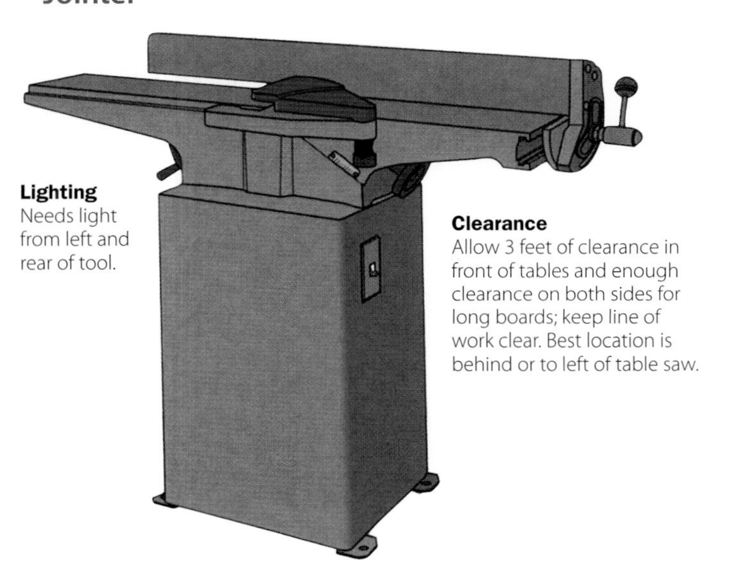

Lighting
Needs light from left and rear of tool.

Clearance
Allow 3 feet of clearance in front of tables and enough clearance on both sides for long boards; keep line of work clear. Best location is behind or to left of table saw.

Lathe

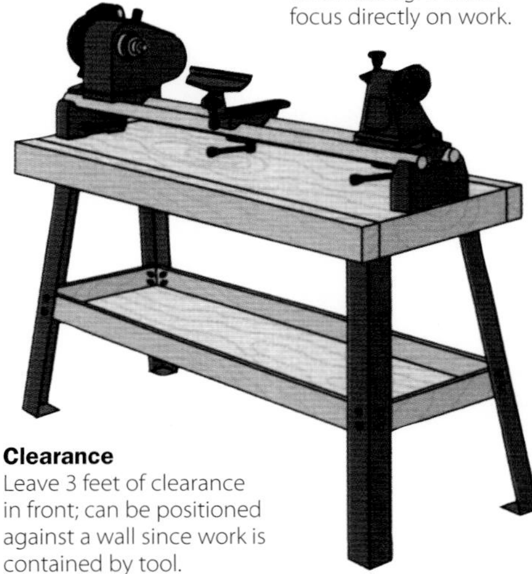

Lighting
Overhead light must focus directly on work.

Clearance
Leave 3 feet of clearance in front; can be positioned against a wall since work is contained by tool.

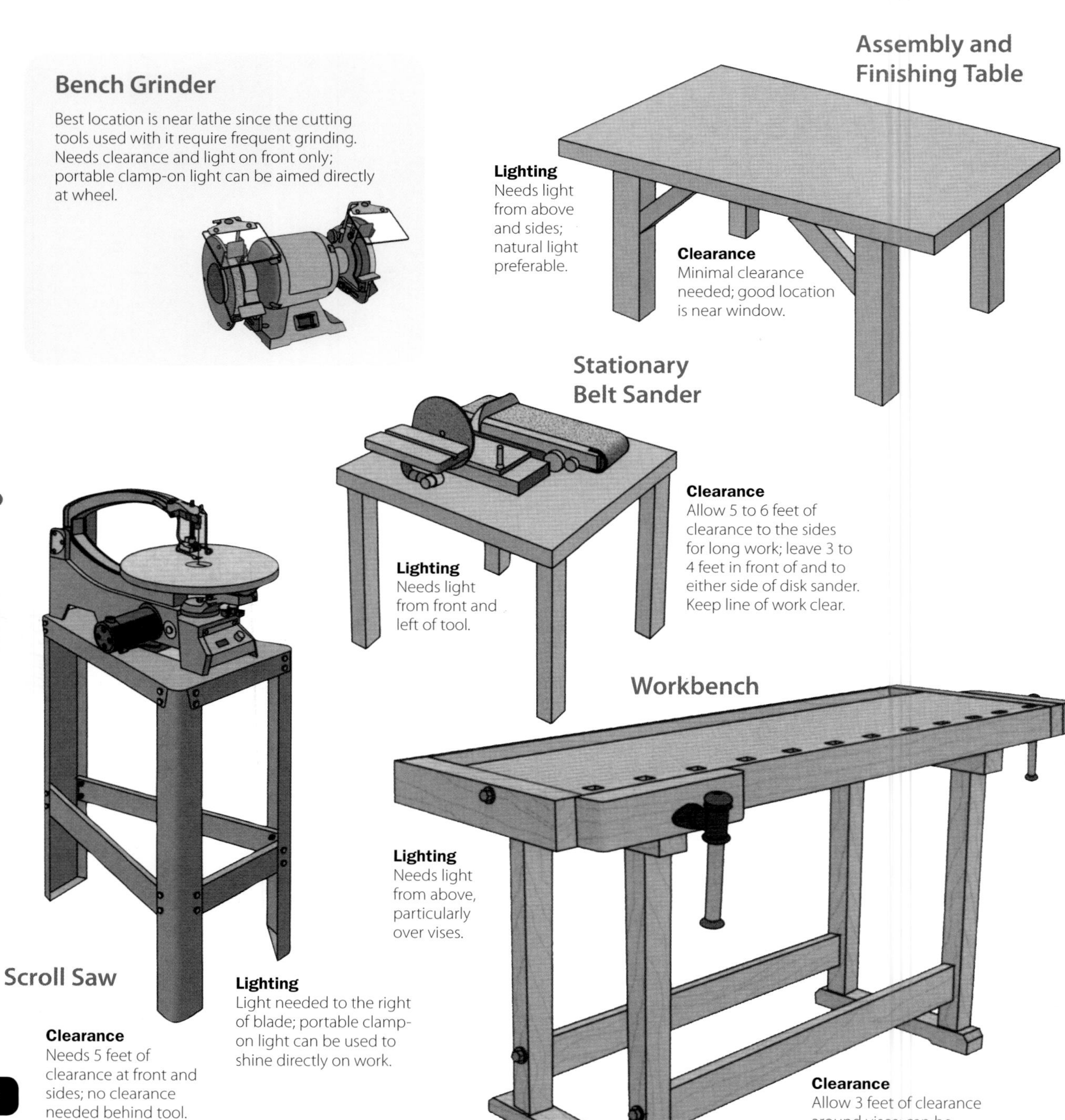
Bench Grinder
Best location is near lathe since the cutting tools used with it require frequent grinding. Needs clearance and light on front only; portable clamp-on light can be aimed directly at wheel.
Assembly and Finishing Table
Lighting
Needs light from above and sides; natural light preferable.
Clearance
Minimal clearance needed; good location is near window.
Stationary Belt Sander
Clearance
Allow 5 to 6 feet of clearance to the sides for long work; leave 3 to 4 feet in front of and to either side of disk sander. Keep line of work clear.
Lighting
Needs light from front and left of tool.
Workbench
Lighting
Needs light from above, particularly over vises.
Clearance
Allow 3 feet of clearance around vises; can be placed against wall if necessary.
Scroll Saw
Lighting
Light needed to the right of blade; portable clamp-on light can be used to shine directly on work.
Clearance
Needs 5 feet of clearance at front and sides; no clearance needed behind tool. Good location is against wall.

Scale Drawings of Stationary Tools

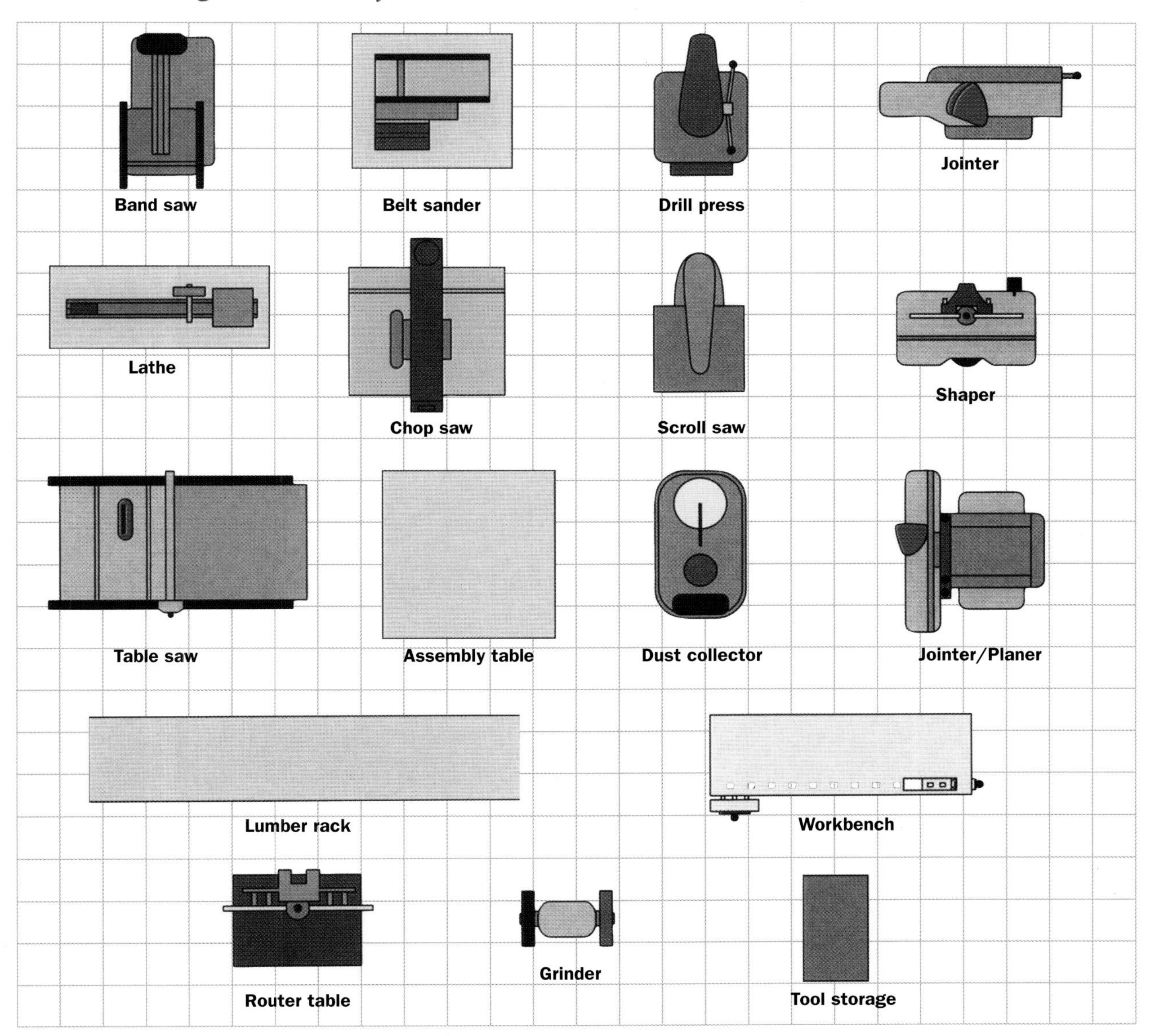

Laying out a workshop on paper
The illustrations above are overhead views of a dozen typical stationary tools drawn at a scale of ¼ inch to 1 foot. To facilitate the task of arranging your tools on the shop floor, sketch your workshop space on a sheet of similarly scaled graph paper. Then photocopy this page, cut out the tools you need, and arrange the cutouts on the grid to determine the best layout for your shop. Consider the space and light requirements of the tools *(pages 16–18)* when assigning space to each one. Also factor in your shop's electrical and lighting needs *(pages 25–27)*. Use the sample layouts of a small-, medium-, and large-size shop beginning on page 20 as guidelines to get you started.

Shop Organization

Layout of a Small Shop

The illustration below shows one way of making efficient use of the space in a small shop—in this case, one-half of a two-car garage. The three stationary machines chosen are essential for most projects: the table saw, the jointer, and the band saw. The saw and jointer are mounted on casters so they can be moved if necessary. With the bench and table there is ample space for hand tool and portable power tool work. The storage space—perforated hardboard and shelving—is located along the walls; a lumber rack is positioned near the garage door. Any exposed framing in the ceiling could also be used to hold stock. Refer to the key in the bottom right-hand corner of the illustration for the type and location of electrical outlets and light fixtures. Note that there is an overhead master switch (near the bench's tail vise) that controls all three machines. Attention is also paid to feed direction of each machine (represented by the arrowhead in the key); the access door to the shop is always in the user's field of vision. **Caution:** If your shop shares space with motorized equipment you will not be able to spray finishes.

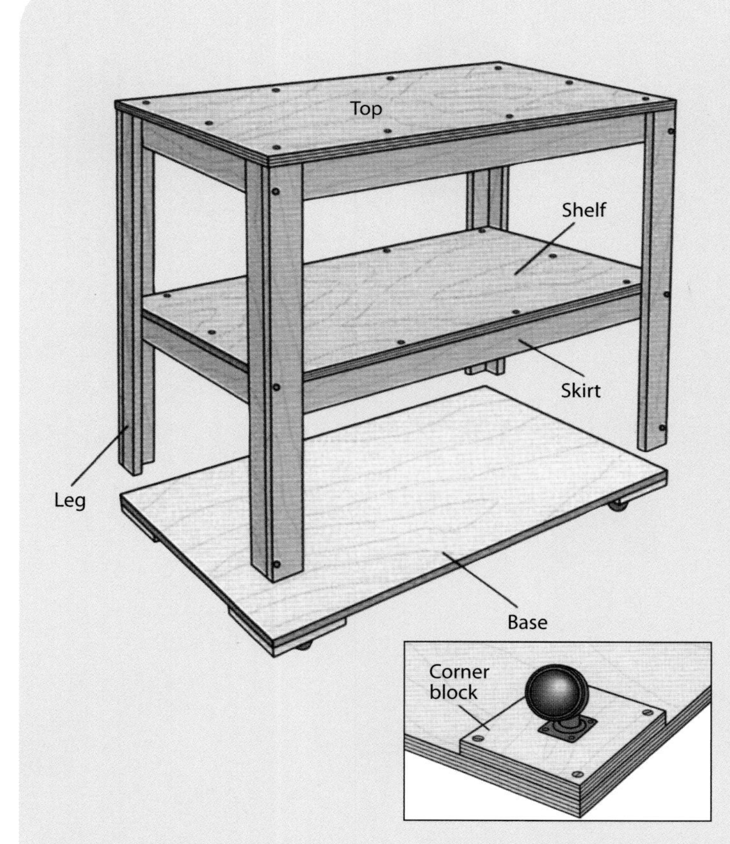

A Shop Dolly

To wheel workpieces or large projects around the shop, use the shop-built dolly shown below. Start with the base and corner blocks, cutting them from ¾-inch plywood to a size that suits your needs. Screw the corner blocks in place, then fasten a caster onto each block (above). To build the shelved section, cut the skirts and the eight pieces for the legs from 1-by-3 stock; the shelf from ½-inch plywood; and the top from ¾-inch plywood. The legs should be long enough for the top to sit at a comfortable height. Screw the leg pieces together, then attach the skirts to the legs' inside faces. Fasten the shelf and the top to the skirts. Secure the legs to the base with angle brackets.

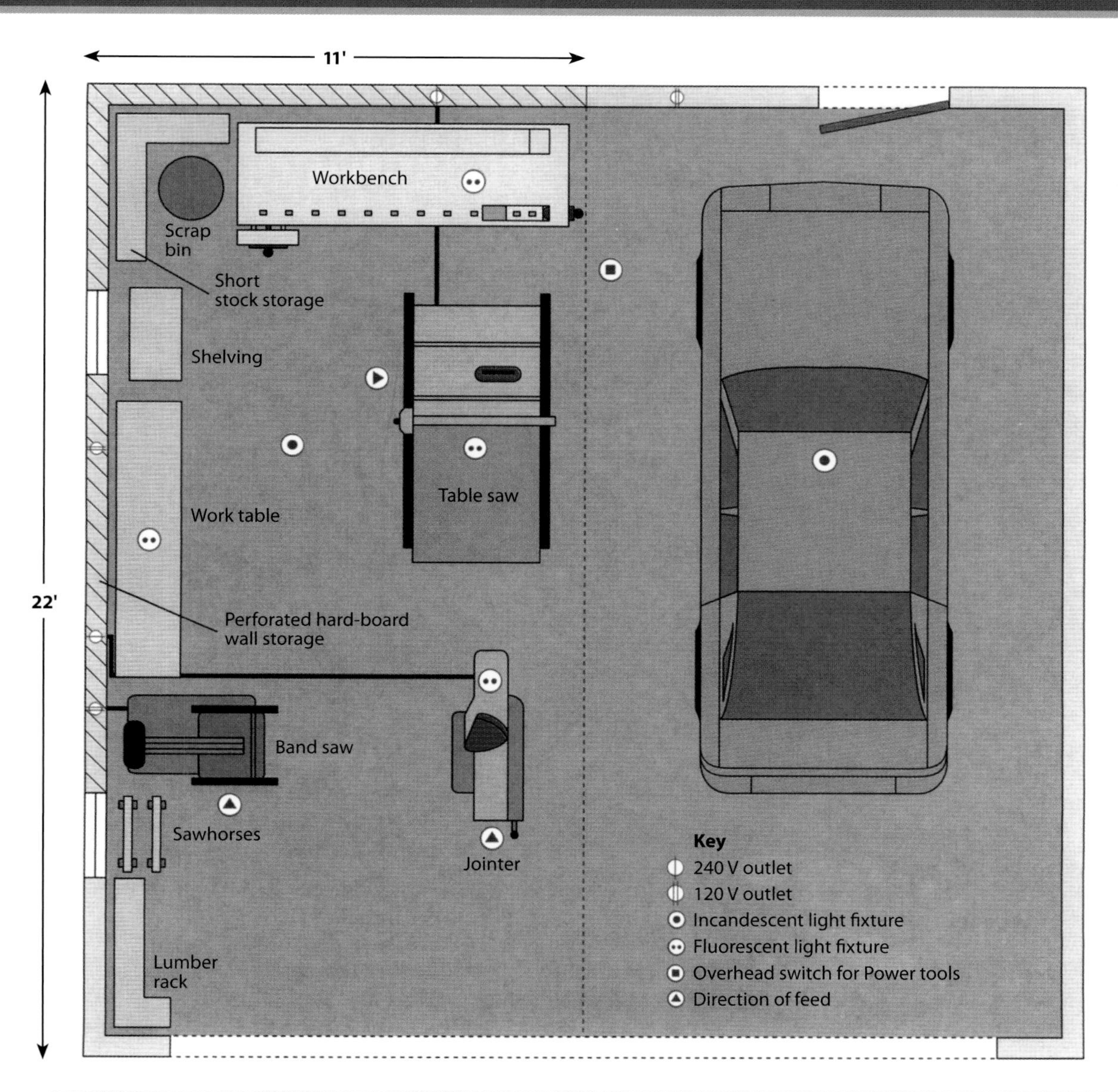

Shop Tip

A Table Saw on Wheels

Because it is the largest and heaviest woodworking tool in many shops, a table saw usually stays put, which can be a drawback in a small shop where space is at a premium. By mounting it on wheels, however, you can easily shift your saw out of the way when it is not in use. If your saw did not come with a wheeled base, measure the base of the motor housing and have a metalworking shop build a rolling base to your specifications. For maximum maneuverability, the base should have three wheels, including one that pivots. Keep the saw from moving or tipping when it is in use by wedging two triangular wood shims under the wheels at the front of the base.

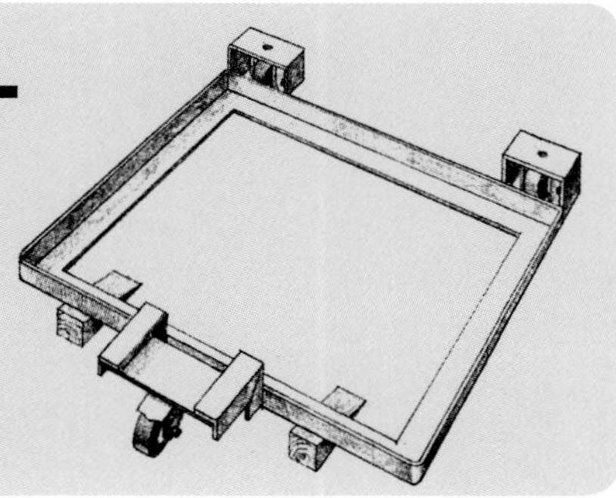

Saving Space

Setting up a shop in the attic

Attic shops have several strikes against them: They are often uninsulated and their floors are not designed to support heavy weight. In addition, headroom is limited and access can present problems, especially if you are working with long planks or full sheets of plywood. But for a luthier, carver, or woodworker who specializes in small projects, an attic can be an ideal spot for a shop. As shown in the illustration at left, nailing sheets of sheathing-grade plywood to the joists will produce a floor that is sufficiently sturdy to hold up a workbench and one of the lighter stationary machines, like the band saw. The spaces between the studs and rafters and down near the eaves—where the roof and attic floor meet—are ideal for storing lumber, tools, and supplies.

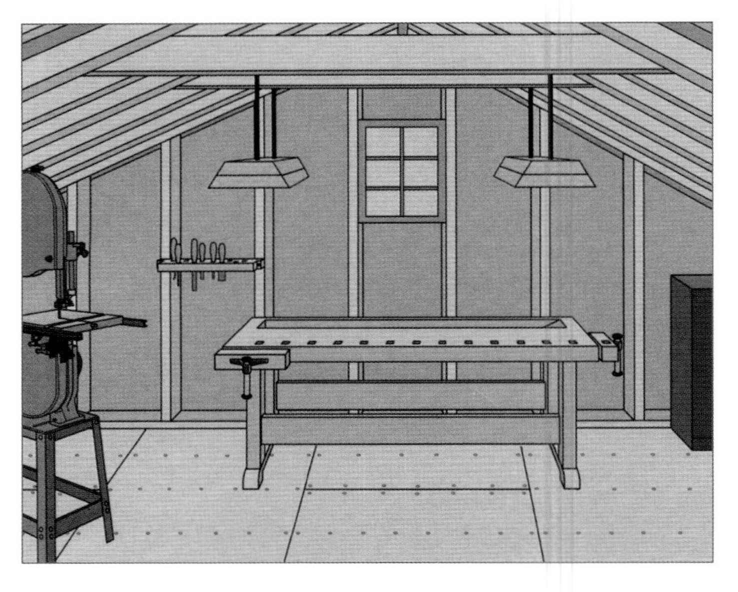

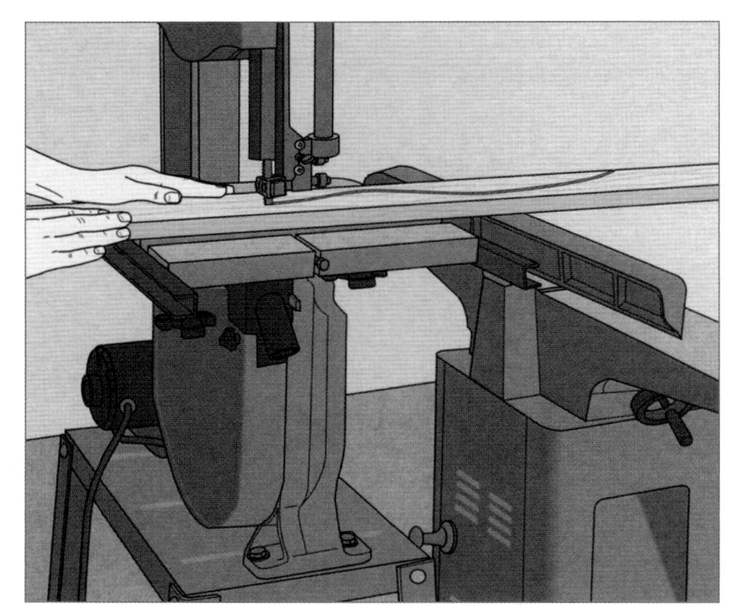

Positioning stationary machines in a confined space

If your workshop is cramped you may have to forego an ideal placement of stationary machines to allow you to make the most of your limited space. Consider the design of your machines and the feed direction you need to use; you may be able to place two machines close together if they are matched properly. The high table of a band saw and the feed direction normally used with the machine, for example, makes it an ideal match in a tight space with a jointer *(right)*. The two can be placed close together while still providing adequate space to operate each machine at separate times.

Layout of a Medium-Size Shop

Setting up a basement shop

In the medium-size shop represented below, the machines are positioned so users will see the door near the bench; perforated hardboard and shelves for storage line the perimeter of the shop (supplies can also be stored under the stairs); and the lumber rack is located near the main access door at the foot of the stairs. This shop has room for a lathe, a drill press, and a dust collector. A work table for glue-up and finishing is positioned at a window with an exhaust fan. The table saw is equidistant from the stock preparation area in front of it, the workbench to one side, and the work table behind it.

Key

- 240 V outlet
- 120 V outlet
- Incandescent light fixture
- Fluorescent light fixture
- Overhead switch for power tools
- Direction of feed

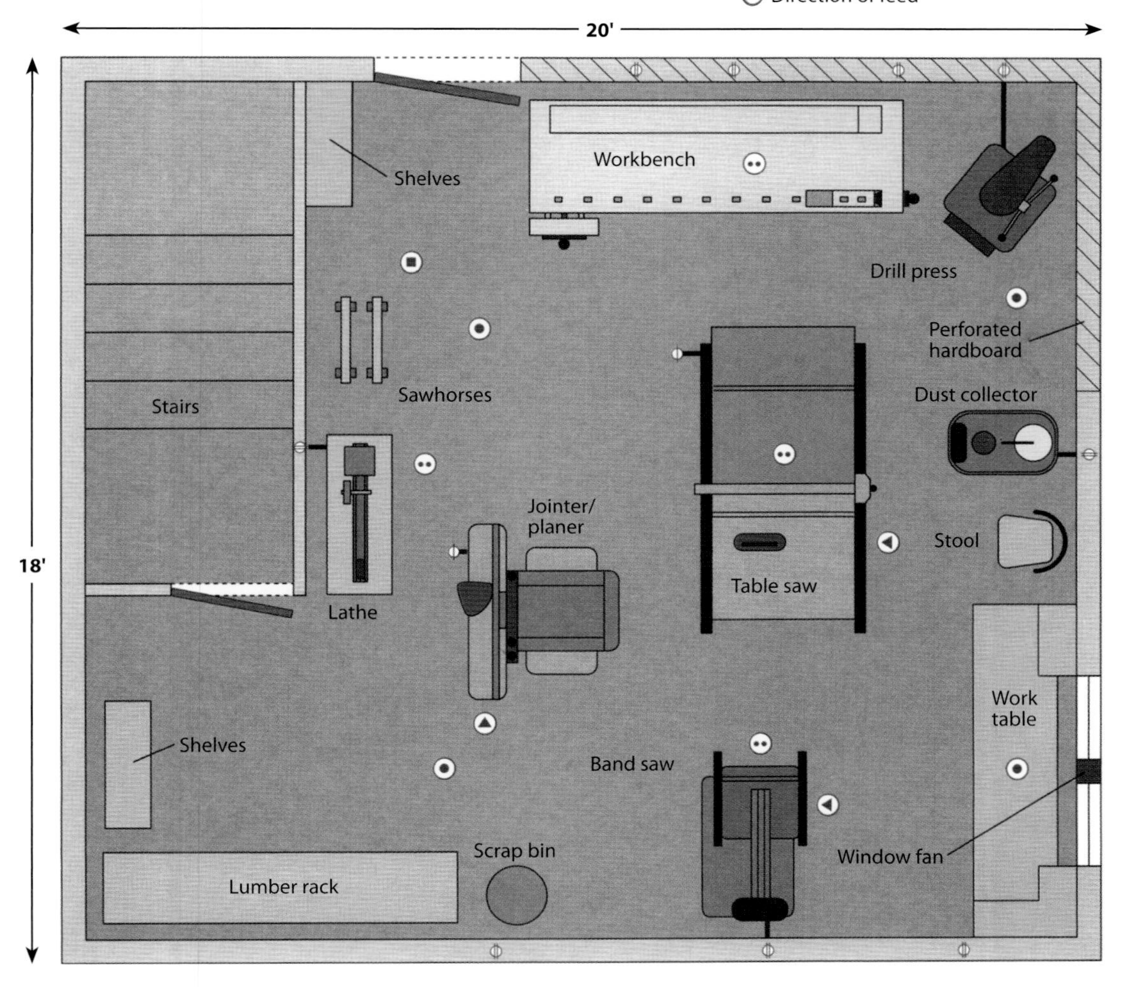

Layout of a Large Shop

Converting a two-car garage

Setting up a shop with all the features shown below calls for a large space, like a two-car garage. This shop has many of the characteristics of the smaller shops examined earlier, with additional tools and conveniences that allow it to handle a wider range of projects. At one corner is a spacious finishing room, partitioned from the rest of the shop and equipped with an explosion-proof fan to exhaust fumes. The shop includes a bathroom with a sink and a toilet. In addition to the machines featured earlier, this has a radial arm saw, shaper, and planer. The shop boasts three separate work surfaces: one in the finishing room, one for glue-up near the drill press, and a workbench beside the table saw. A shop of this size would need an independent electrical service panel to power all the tools. To keep the wiring out of the way, half the floor is covered with a raised ¾-inch plywood floor; as shown on page 28, an understructure of 1-by-2s is laid on the concrete floor on 12-inch centers and the plywood is nailed to the boards. Wires are run in conduits under the plywood between the 1-by-2s.

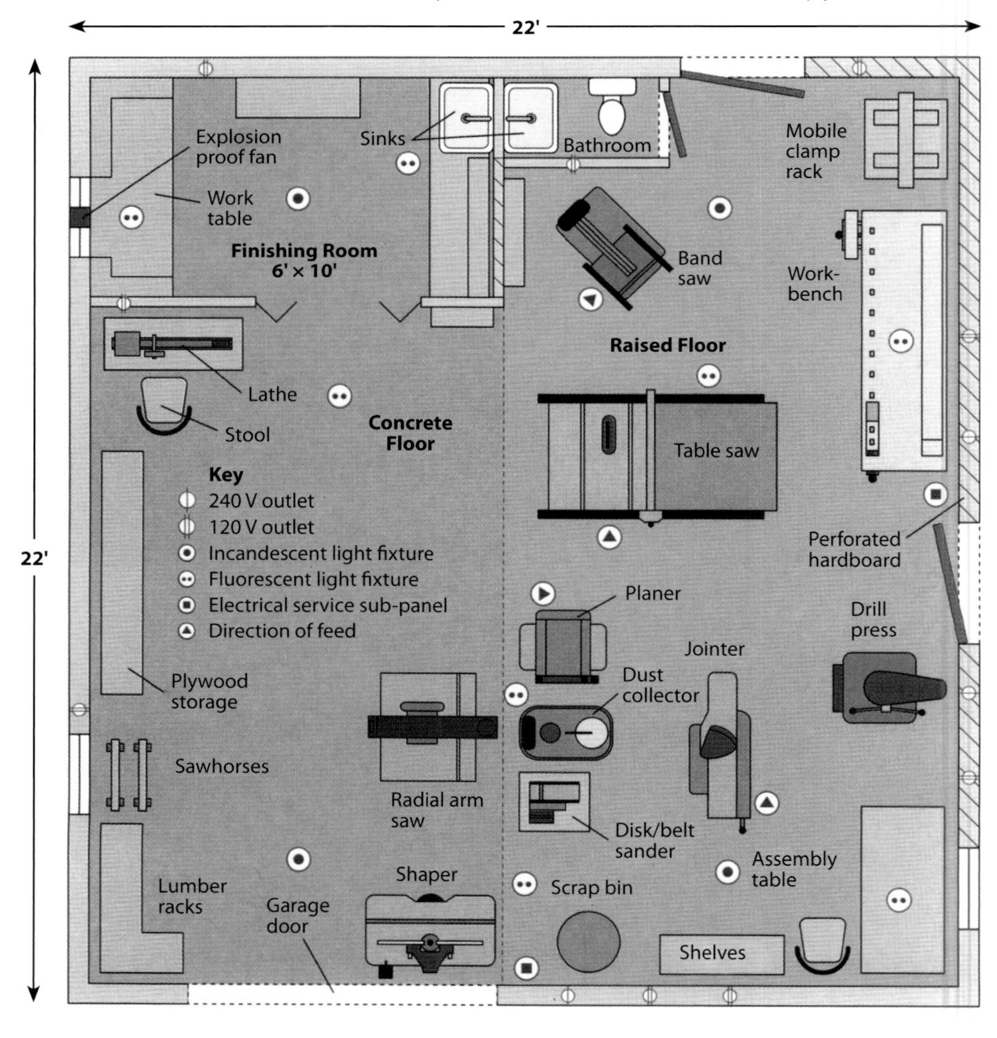

Lighting

If you find yourself cutting off line or cannot properly examine a finish unless you take your work outside, the lighting in your workshop may need an upgrade. At best, a poorly lit shop will merely bring on fatigue; at worst, it can contribute to sloppy, imprecise work and to accidents.

Fluorescent lights are the most popular type of workshop lighting fixture. They cast a relatively shadowless light, the tubes are long-lasting, and they use 20 percent to 30 percent less electricity than incandescent lights of the same brightness. Many woodworkers find that too much fluorescent light can result in fatigue and headaches, however, and prefer the warmth of incandescent and tungsten lights.

At a minimum, a shop bigger than 120 square feet needs 2 watts of incandescent light or ¾ watt of fluorescent light per square foot. As in the electrical layout illustrated on page 27, shop lights should be circuits separate from your tools. Ideally, the light fixtures will be divided between two separate circuits. As a rule of thumb, do not exceed 1600 watts on one 20-amp circuit. Also, distribute lighting fixtures around the shop; mounting a single fixture in the middle of the ceiling will make it difficult to illuminate the shadowy areas at the edges of the shop.

If possible, make the most of natural light; there is no better substitute, especially for hand-tool work and finishing. Trying to evaluate planing, sanding, and finishing jobs under artificial light can be frustrating. Both fluorescent and incandescent light tend to distort or disguise the surface texture of natural and finished wood surfaces. Natural light, particularly from the north, has a soft, non-glare quality. If your shop has a window that faces north, place your workbench under it.

Keep in mind that upgrading the lighting in your shop need not entail purchasing expensive fixtures and rewiring the system. Simply painting a concrete floor a light color or covering the ceiling with white tiles will allow these surfaces to reflect light, rather than absorb it.

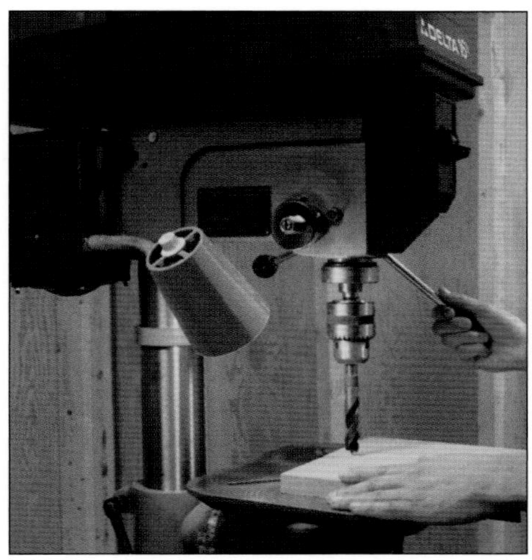

A clamp-on lamp can shed all the light you need to work safely at a tool. Mounted on a drill press, this lamp's flexible neck aims a 40-watt bulb directly at the machine's work table

Shop Tip

A bench-dog lamp support

For a movable source of light at your workbench, attach a desk lamp to one of the bench dogs. Bore a hole the same diameter as the shaft of the lamp into the head of a wooden dog *(page 38)*. The light can then be positioned at any of the dog holes along the bench.

Electrical Power

Electric power requirement should be considered early in the process of planning a shop's layout. Allow for growth. Then, as you add new tools and light fixtures, you will avoid the headaches of an inadequate system: repeated tripping of circuit breakers or blowing of fuses, and octopus adapters funneling several power cords into one outlet.

If you plan to wire your shop to your home's main service panel, be sure that your electrical supply has enough additional power. You can get a rough idea of how many amperes your shop will draw from the system by totaling the amperage of all the tools you plan to use and dividing the result in half. If your system is barely able to handle the demands being placed on it by your household, you probably will need to upgrade your service entrance—in other words, increase the number of amps the service panel can draw from the utility company. If the shop will be some distance from the main service panel, it is a good idea to install a 50-amp sub-panel dedicated to the shop. Another point to remember: Any woodworking machine that draws more than six amps should be on a separate (dedicated) circuit, unless the tool's motor is shielded.

Refer to the illustration on page 27 as a guide to planning the electrical layout of your shop. As you plan, remember that even simple electrical jobs, like extending a circuit or replacing an outlet, can be dangerous. They can also cause a fair amount of damage—ranging from burned-out tool motors to a house fire—if they are carried out improperly. Unless you are qualified and comfortable with the idea of wiring your shop to the electrical system, have a qualified electrician do the job.

Electrical Layout Tips

- When planning the electrical layout for your shop, make sure that outlets for power tools and lighting fixtures are on separate circuits.
- Unless your shop has bright windows or your lights are equipped with battery backups, include at least two separate lighting circuits in your electrical layout. In the event one circuit is disabled, the lights plugged into the other circuit will still work.
- Place outlets close to the eventual location of the tools they will power; distribute outlets all around the shop to allow for future tool acquisitions.
- Avoid locating outlets on the floor; they will eventually become filled with sawdust and be a fire hazard.
- Avoid plugging tools into one outlet using an octopus adapter; this can overload your electrical system, and is a sign that the wiring of your shop is inadequate. Upgrade the system by installing new outlets and wiring them to a separate circuit on the service panel.
- Protect any new outlet in a garage or basement by installing a ground-fault circuit interrupter (GFCI).
- Never work on the wiring of the service panel; entrance wires may remain live even when power is shut off at the main circuit breaker or fuse block.
- Make certain that any new circuits or service sub-panels installed in your home or outbuilding are grounded to the main service panel. Individual outlets must also be grounded.
- Do not take off the cover of the service panel.
- Never work on your wiring in damp or wet conditions.
- Do not touch a metal faucet, pipe, appliance, or other object when working on your wiring.
- Never splice a power cord or an extension cord, or remove the grounding prong from a three-prong plug.
- Use an extension cord to supply electricity to an area only temporarily—not as permanent wiring.
- Never run a power cord or an extension cord under a rug, mat, or carpet; do not fasten the cord using tacks, pins, or staples.
- Never replace a blown fuse with one of higher amperage; do not use a penny, a washer, or foil as a substitute for a fuse.
- If a circuit breaker trips or a fuse blows repeatedly, check for a short circuit, and determine whether the circuit is overloaded.

Electrical Layout for a Medium-Size Shop

Wiring the shop

The illustration at left shows one electrical layout for a medium-size shop. The shop has six separate electrical circuits: four for tools and two for lighting. The basic principle to keep in mind is that no circuit using 12-gauge wire should carry more than 80 percent of its capacity; for 20-amp circuits, this means the combined amperage of the tools on the same circuit plus 25 percent of the rating of the largest motor must not exceed 16 amps. In this shop, the table saw and jointer are on separate 240 V circuits; their power cords are suspended from the ceiling with twist-type outlets, which keep the plugs in place. With a combined load of 15 amps, the band saw and the drill press are on the same 120 V circuit; the lathe is on another. Additional outlets on the 120 V circuits can be used for portable tools. The incandescent and fluorescent lighting circuits are separate so that if one fails the other will still work.

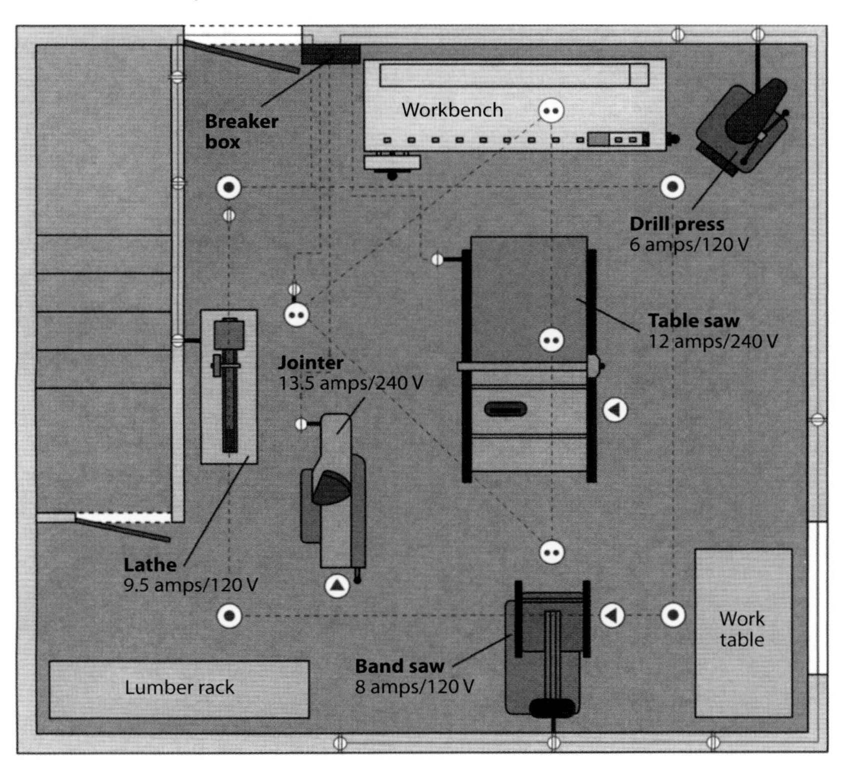

Key

- 240 V outlet
- 120 V outlet
- Incandescent lighting fixture
- Fluorescent lighting fixture
- Direction of feed

Shop Tip

Power cord covers

Power cords lying loosely on a shop floor are accidents waiting to happen. If your shop does not have overhead outlets for your machines, cover the cords with wood bridges. Cut the covers from ¾-by-2½ hardwood stock and rout a groove along the length of one face to house the cord. Cut bevels on the opposite face so the cover will not be an obstruction.

Floors, Walls, and Ceilings

Since most workshops are set up in basements or garages, concrete floors are a common feature. Yet for anyone who has to spend much time standing on concrete or sweeping it clean, the material can prove both uncomfortable and inconvenient. The hard surface is particularly tough on tools that are dropped accidentally.

Simply painting a concrete floor with a paint made specifically for the purpose will keep down the dust and make the surface easier to clean. Adhesive vinyl floor tile can be laid down as well. Yet many woodworkers prefer the comfort of a raised wooden floor. A simple floor can be constructed from sheets of ¾-inch plywood laid atop a grid of 1-by-2s on 12-inch centers. Not only is this type of floor easier on the feet, but wiring for stationary power tools can be routed underneath the raised surface in ½-inch plastic or steel conduit.

Unlike the walls of most homes, those of separate workshops seldom are insulated. If you live in a northern climate, you can increase the thermal efficiency of your shop by covering its walls with wood paneling or sheet material, and filling the gap in between studs with insulation. Wood paneling in particular creates a warm, comfortable atmosphere. Interior wall covering will make your shop quieter too, since the walls will absorb some of the din of your power tools. As a bonus, you can conceal wiring behind the walls. Make sure the basement walls do not leak before covering them with insulation and paneling.

To hide the exposed joists, ducts, and wiring above your head, consider installing a ceiling. A suspended tile ceiling, in which the tiles sit in a framework of supports hanging from the joists, is one popular option. In a large shop, a dropped ceiling such as this will also help retain heat. Acoustical ceiling tiles are an inexpensive alternative; the tiles are attached to furring strips that are nailed to the joists.

Standing in one place for hours on a concrete floor can strain your feet and legs. An old piece of carpet or a commercial anti-fatigue mat provides a cushion that can he easily moved about the shop

Shop Tip

Making the transition to a raised floor

If part of your shop has a raised floor, you can make a smooth transition from the lower concrete floor with several beveled 2-by-6s laid end-to-end. Cut a rabbet in one edge of each 2-by-6 to accommodate the plywood floor and the 1-by-2 grid underneath. Then bevel the opposite edge, forming a ramp to facilitate moving items from one floor to the other. Nail or screw the plywood to the 2-by-6s.

Heating and Ventilation

Heating is a necessity for most shops in North America. Some woodworking tasks demand it; gluing and finishing in particular require steady temperatures. Heating your shop also makes it more comfortable and safe; numb fingers invite accidents.

If your shop is some distance from your home's furnace, a separate heating system will be needed. Many woodworkers swear by wood heat; it has the added benefit of consuming scrap pieces. Yet this means frequently feeding the stove and cleaning the chimney; insuring your shop against fire can also be a problem. Electric baseboard units are more convenient, but can contribute to high utility bills and frequently are clogged with sawdust.

Portable kerosene and propane burners should be avoided in the shop, since they use an open flame and emit toxic exhaust. Coil-type electric heaters are also a fire hazard.

Whichever heating system you choose, keep the area around it free of sawdust and place it away from the finishing and wood storage areas. And remember, any system will be improved by good ventilation.

Consider your need to control humidity. In shops in humid climates, too much moisture means an investment in a dehumidifier to keep wood dry and tools from rusting. Shops in more arid climates face the opposite dilemma and may require a humidifier.

Finally, every shop requires adequate ventilation. Airborne sawdust and toxic finishing vapors may not be as visible a danger as kickback on a table saw but the threat they pose is just as real. While fire or explosions due to high concentrations of sawdust or finishing vapors are rare, they can be devastating. A good ventilation system changes the air often enough to maintain safe levels of airborne dust and fumes. It should include dust collection equipment at each stationary power tool that produces sawdust *(page 54)*, and a general exhaust setup (below) to remove the dust and fumes that remain.

While window fans or bathroom-type vent models are fine for general exhaust purposes, a finishing booth or spray room requires something different: An explosion-proof tube-axial fan is recommended. Fans are rated by the amount of air that they move, measured in cubic feet per minute (cfm). Divide the cubic volume of your shop (its length times its width times its height) by 6 to find the rating needed to change the air 10 times per hour—the minimum level for safe ventilation.

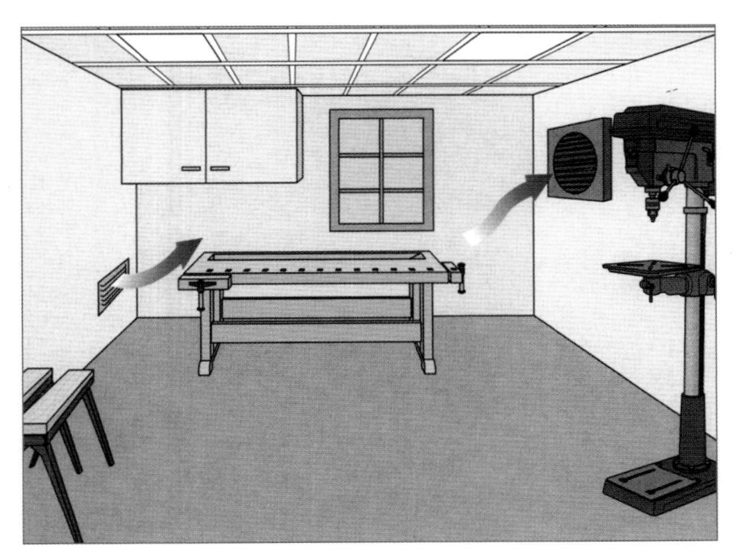

Ventilating a Shop

Installing a general exhaust setup

If your shop does not have windows or doors to provide proper cross-ventilation, install an exhaust setup to clean the air. The system shown at left is a simple one, consisting of an air intake at one end of the shop connected to the outdoors or your home's air ducts, and an explosion-proof fan mounted in the wall at the opposite end. The intake is covered with a furnace or air-conditioning filter to clean the incoming air. The exhaust fan is placed higher than the intake, causing the air that rises to be drawn out of the shop. For best results, orient the exhaust setup along the longest axis of your shop.

Workbench

The workbench is the cornerstone of the woodshop, with a history almost as old as woodworking itself. Examples of primitive workbenches have been found dating back more than 2,000 years. Woodworkers in ancient Rome advanced the basic design, devising benches with simple stops that allowed them to secure pieces of wood. Until that time, craftsmen were forced to hold their work, cutting or shaping it with one hand while chopping or planing with the other. Further improvements came slowly, however, and vises were only added centuries later.

With each refinement the workbench has assumed an increasingly indispensable role in the workshop. It is little surprise that many call the workbench the most important tool a woodworker can own.

A good workbench does not take an active role in the woodworking process—it does not cut wood or shape it—but the bench and its accoutrements perform another essential task: They free your hands and position the work so you can cut, drill, shape, and finish efficiently. In the past, even the most-used benches have fallen short of the ideal. With its massive, single-plank top, the Roubo Bench of the 18th Century was popular throughout Europe, yet it had no tail vise or bench dogs to hold a workpiece; instead, the task was done by a system of iron holdfasts and an optional leg vise. One hundred years later, the American Shakers improved on the Roubo. Their bench was a large affair that sported a laminated top, a system of bench dog holes, an L-shaped tail vise, and a leg vise. The Shaker bench was not too different from the modern cabinetmaker's bench pictured at right.

The design of the workbench has changed little since the early 19th Century; only its accessories and manner of assembly have been altered. In fact, some claim that the only true innovation has been inventor Ron Hickman's ubiquitous Workmate™. Developed in the 1960s, the Workmate™ revolutionized the way many people look at work surfaces, because it provided some of the clamping abilities of a standard workbench with a collapsible, portable design.

Although the Workmate™ has found a niche in workshops around the world, many woodworkers—both amateur and professional—still opt for nothing less than a solid maple or beech bench. Often they choose to build their own, believing that the care and attention paid in crafting such a bench will be reflected in their later work. The chapter that follows shows how to assemble a modern cabinetmaker's workbench, and how to install the vises and accessories needed to turn an ordinary bench into a more flexible work station.

With its origins rooted in an era without power tools, the standard cabinetmaker's bench now incorporates vises designed for use with both power and hand tools.

Anatomy of a Workbench

The workbench shown at right is patterned after a traditional cabinetmaker's bench, and is crafted from solid maple. The bench incorporates two vises considered to be standard equipment: a face vise on the front, left-hand end of the bench, and a tail vise with a sliding dog block mounted on the opposite end.

You can build such a workbench from a kit supplied with materials and instructions. You can buy the plans for a bench and order the materials yourself. Or, you can follow the instructions presented in this chapter and construct a bench to suit your needs. Whichever route you take, a workbench is assembled in three distinct phases: the base *(page 34)*; the top *(page 37)*; and the clamping accessories—vises *(page 40)*, bench dogs, and hold-downs *(page 46)*.

The top surface of most benches is generally between 33 and 36 inches high. The height that is best for you can be determined by measuring the distance between the floor and the inside of your wrist while you stand upright with your arms at your sides.

Finish your workbench with two coats of a penetrating oil-based product, such as tung oil. Not only do these products penetrate the surface and protect the wood, but the finish can be refurbished simply by scrubbing it with steel wool and recoating.

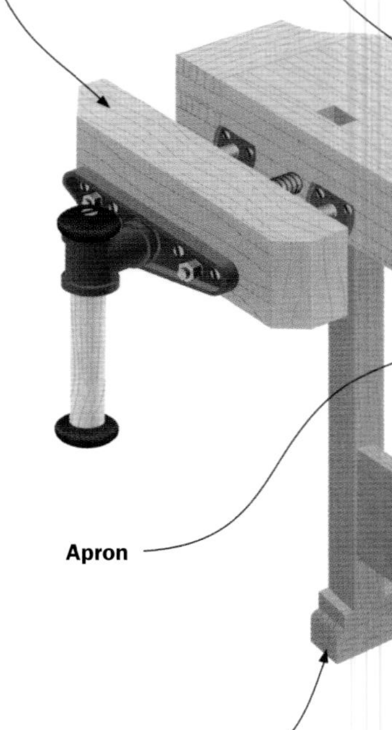

Attaching the end caps of a workbench to the aprons calls for a strong and attractive joinery method. The finger joint (also known as the box joint) and the dovetail joint shown at left are traditional favorites.

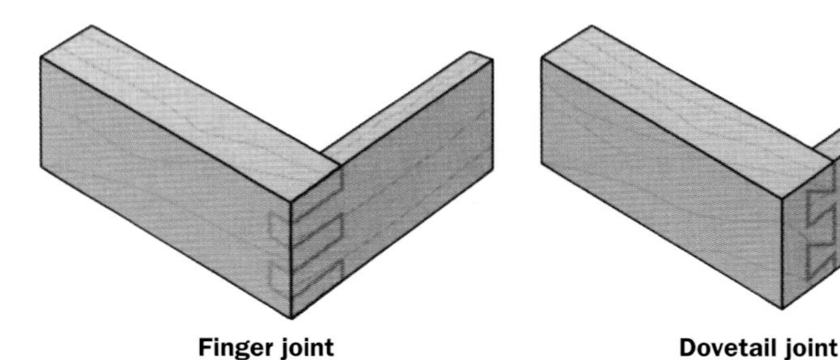

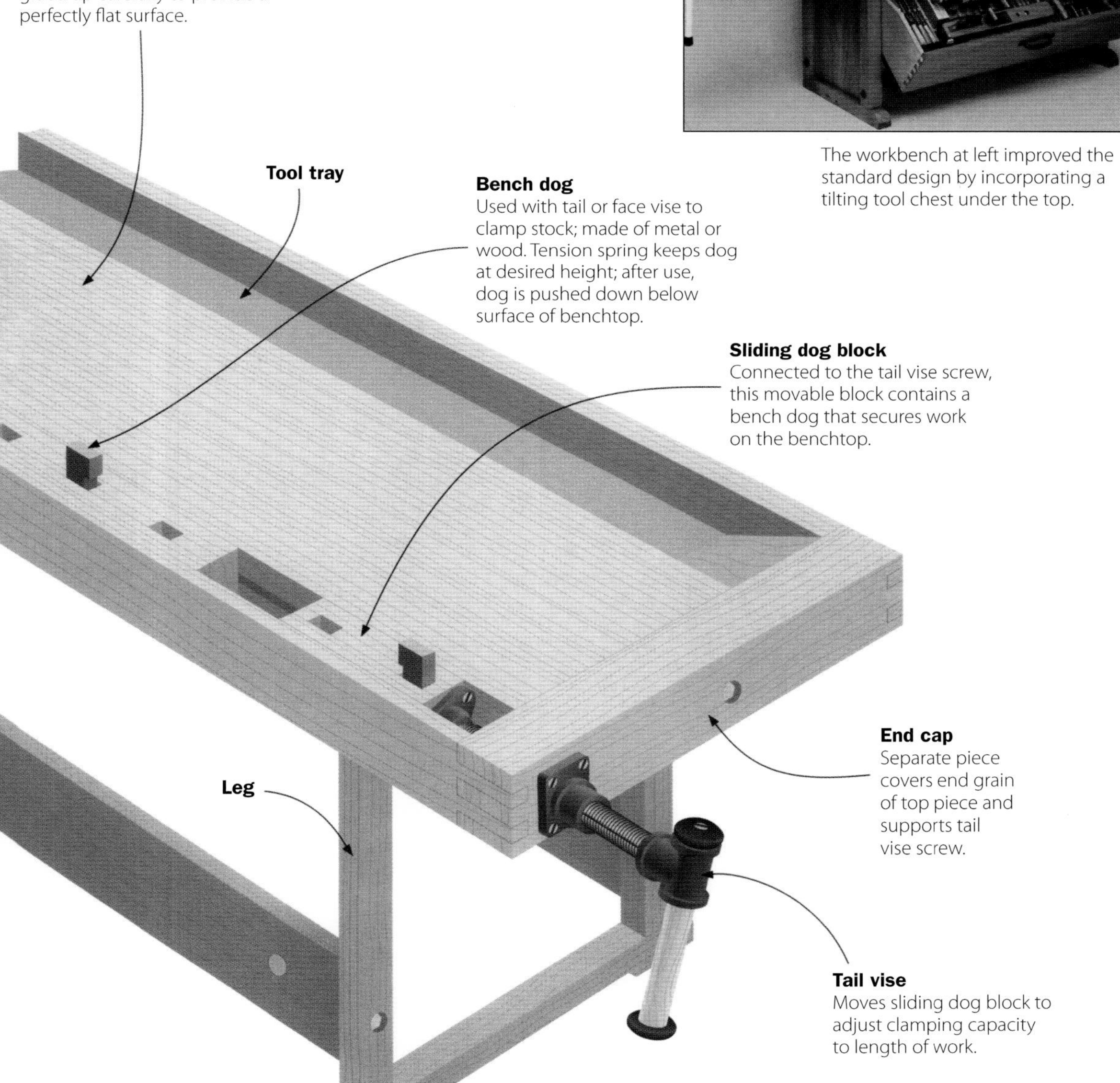

The workbench at left improved the standard design by incorporating a tilting tool chest under the top.

Building the Base

The base of a workbench typically consists of two rectangular frames connected by a pair of stretchers. The frames are essentially identical, each with a foot, an arm, and two legs. The arm of the left-hand frame is sometimes about 3 inches longer than the other arm to provide additional support for the face vise.

For a bench like the one shown on page 31, use 8/4 maple (1¾ inches thick after surfacing). The feet, arms and legs are made from two boards apiece face-glued together, and then reduced to the proper thickness on the jointer and planer. If you wish to build the base with mortise-and-tenons, cut four-shouldered tenons at the end of the legs and rout matching mortises in the feet and arms. Tenons are also cut at the ends of the stretchers with mortises required in the legs. The illustration below shows a knockdown alternative to assembling the base with mortise-and-tenons.

The joints between the stretchers and the legs need to be solid, yet sufficiently flexible to be taken apart should you want to move the bench. Consequently, knockdown hardware designed for the purpose is often used to join the stretchers to the legs. The pages that follow detail some other methods of reinforcing knockdown connections.

Butt joints connecting the legs of a workbench to the stretchers can be reinforced with hardwood knockdown fittings. The fittings are inserted into mortises cut into the ends of the stretchers; matching machine bolts and nuts are then used to secure the joint.

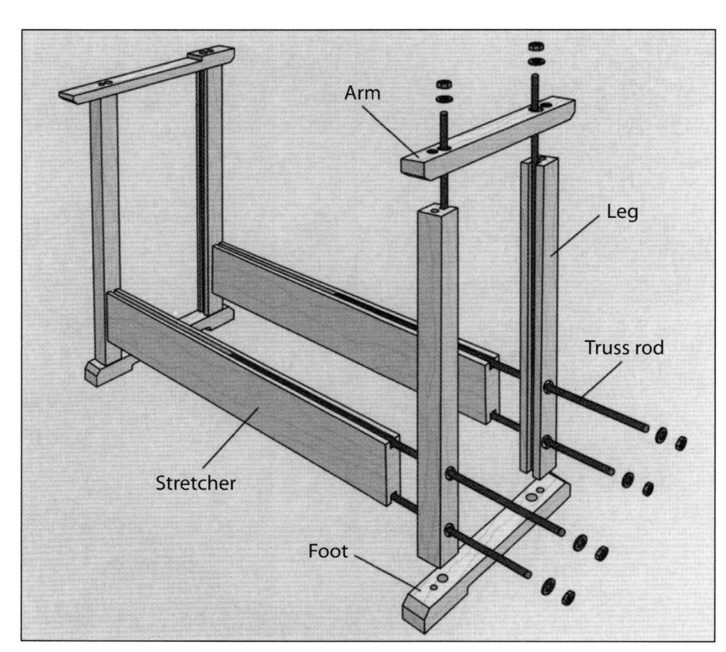

Reinforcing Knockdown Joinery

Using truss rods

Instead of using mortise-and-tenon joints to build the base, use butt joints reinforced by truss rods, as shown at right. Available in kits, the rods can be loosened or tightened after assembly to compensate for wood movement as a result of changes in humidity. Rout grooves for the rods into the edges of the stretchers and the inside edges of the legs; the depth and width of the channels should equal the rod's diameter. Test-assemble the base and mark the groove locations on the legs and arms. Then bore a hole at each mark, making the diameter equal to that of the rods; countersink the holes so you can drive the nuts flush with the wood surface. Assemble the base, fitting the rods into the grooves and holes, and tightening the connections with washers and nuts. Cover the grooves with solid wood inlay if you wish to conceal the rods.

Using machine bolts and wood blocks
To reinforce the connection between the legs and stretchers, glue a wood block of the same thickness as the stock to each edge of the stretchers. The blocks will increase the contact area between the stretchers and the legs. Once the glue is dry, cut a tenon at the end of each stretcher and a matching mortise in the leg. Fit the pieces together and bore two holes for machine bolts through the leg and the tenon in the blocks; countersink the holes. Make the connection fast by fitting the bolts into the holes, slipping on washers and tightening the nuts *(right)*.

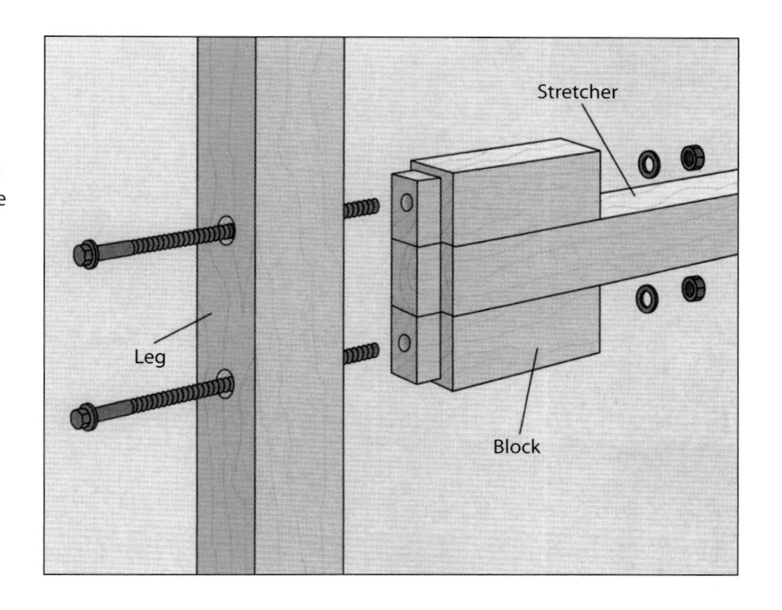

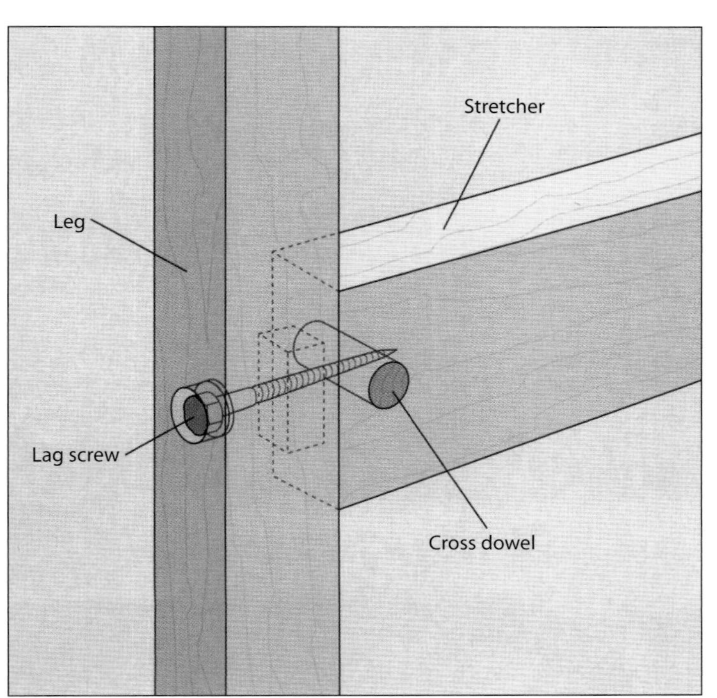

Using lag screws and dowels
Another way to strengthen a mortise-and-tenon joint between the stretchers and legs is shown at left. Cut a 1-inch-diameter hardwood dowel to a length equal to the thickness of the stretcher. Then bore a 1-inch-diameter hole through the stretcher about 1½ inches from its end. Also bore a hole for a lag screw through the leg, stopping the drill when the bit reaches the hole in the stretcher; countersink the hole so the screw head will sit flush with the surface. Fit the stretcher tenon into the leg mortise, tap the dowel into place in the stretcher, and drive the screw. Choose a screw that is long enough to bite through the dowel.

Preparing the Feet

Relieving the feet
Once you are satisfied with the fit of the parts of the base, disassemble the stretchers and legs and relieve the feet on the jointer. Install a clamp on the jointer's infeed table to hold the guard out of the way during the operation. Set both the infeed and outfeed tables for a 1/16-inch depth of cut, and clamp stop blocks to both tables to guide the beginning and end of the cut. To make the first pass, lower the foot onto the knives, keeping it flush against the fence and the stop block on the infeed table. Feed the foot across the knives *(right)* until it contacts the stop block on the outfeed table. Keep both hands well above the cutterhead. Make as many passes as necessary to complete the recess, lowering the tables 1/16 inch at a time, and readjusting the stop blocks as necessary.

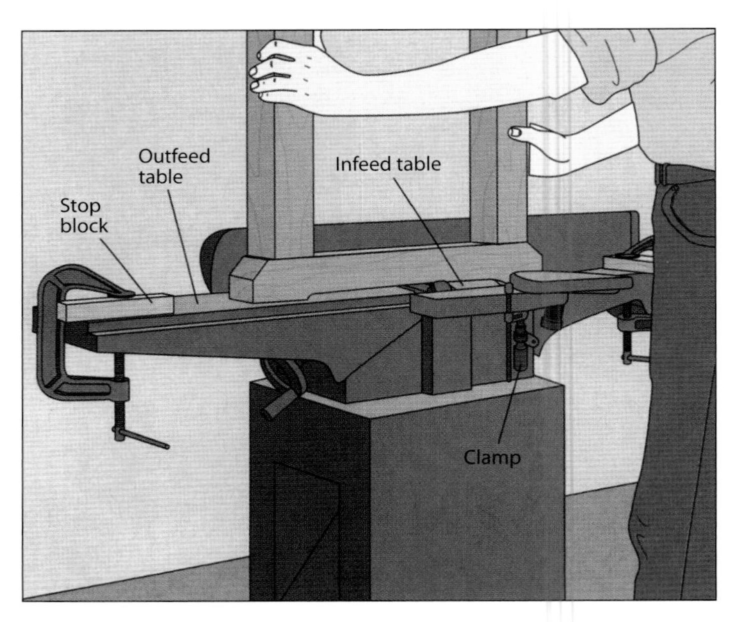

Installing adjustable levelers
To level a workbench on an uneven shop floor, install adjustable levelers in the feet. Each leveler consists of a T-nut and a threaded portion with a plastic tip *(inset)*. Bore two holes into the bottom of the foot near each end. Make the hole's diameter equal to that of the T-nut and its length slightly longer than the threaded section. Tap the T-nuts into the holes and screw in the levelers *(left)*. Once the bench is assembled, adjust the levelers until the benchtop is level.

Building the Top

One of the most important features of a workbench is a perfectly flat top. At one time, a benchtop could be built of solid maple or beech boards 12 inches wide and 2 inches thick. But today such planks are difficult to come by, and benchtop slabs are built up from narrow boards, layers of plywood sandwiched between strips of hardwood, or laminated plywood strips sheathed in hardboard. However, edge gluing solid wood boards together butcher-block style, as shown below, is the time-honored method.

Cut from 8/4 stock, the boards are glued together first, then the slab is cut to length. To minimize warping, arrange the pieces so that the end grain is reversed. Also make sure the face grain of all the boards runs in the same direction. This will make it easier to plane the top surface of the slab smooth.

After gluing up the slab, prepare the dog blocks. They are glued up from a length of 8/4 stock and one of 4/4 stock with the bench holes dadoed out of the thicker board. The sliding dog block for the tail vise is sawn off before the front rail and fixed block are glued together *(page 38)*. Next, the slab, fixed dog block, and rear rail are glued up *(page 39)*; hardwood keys and plywood splines are used to strengthen the connections.

After the sliding dog block, tool tray, and aprons are installed, the final step involves attaching the end caps to the top. Two connections are used. The caps are bolted to the slab and joined to the aprons by means of dovetail or finger joints.

Once the top of a bench is installed on the base, a straightedge held on edge across the surface can be used to check it for flatness.

Anatomy of a Benchtop

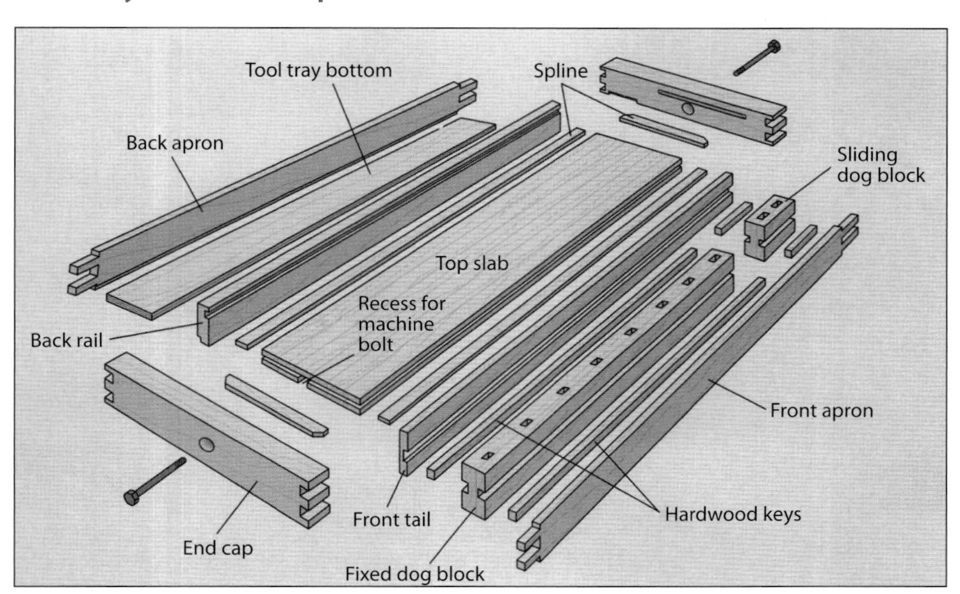

Preparing the Fixed Dog Block

Cutting the bench dog holes
Bench dogs are fabricated from two boards, so it is simple to cut the dog holes in the thicker piece before glue-up. Two steps are involved. First, cut a row of evenly spaced dadoes wide enough to accept the dogs; angle the fixed-block dadoes slightly toward the tail vise, and the sliding-block dadoes away from the tail vise so that the dogs will grip the work firmly when clamping pressure is applied. Next, clamp the board to a work surface and use a chisel to notch the top of each dado to accept the dog heads *(right)*. That way, the dogs can be pushed down flush with the bench surface when they are not in use. Now the two parts can be glued up to form the finished blocks.

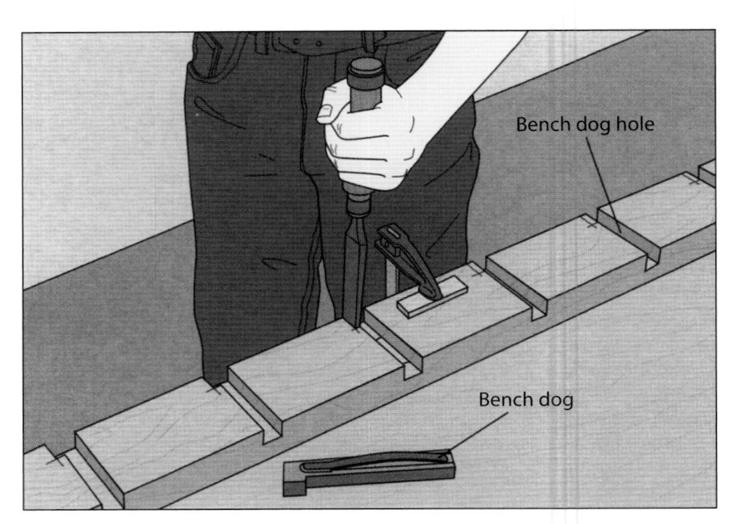

A Jig for Drilling Bench Dog Holes

If you plan to use round bench dogs, you can use the shop-made jig shown below to bore their holes after you glue up the bench top. The jig should be about 10 inches long; the lip is cut from a 1-by-2 and the base from a 1-by-4. After screwing them together, bore guide holes about 8 inches apart and 3 inches from the lip. The holes should accommodate the dogs you will use.

To use the jig, clamp it to the right end of the dog block so the lip is against the front edge and the right-hand guide hole is over the position of the first dog hole. Using the guide holes, bore the first two holes in the bench. For each subsequent hole, remove the clamp and slide the jig to the left until the right-hand guide hole is aligned with the last hole bored. Slip a bench dog through the holes, clamp the jig and bore the left-hand hole *(right)*. Repeat the process until you are finished boring all the holes.

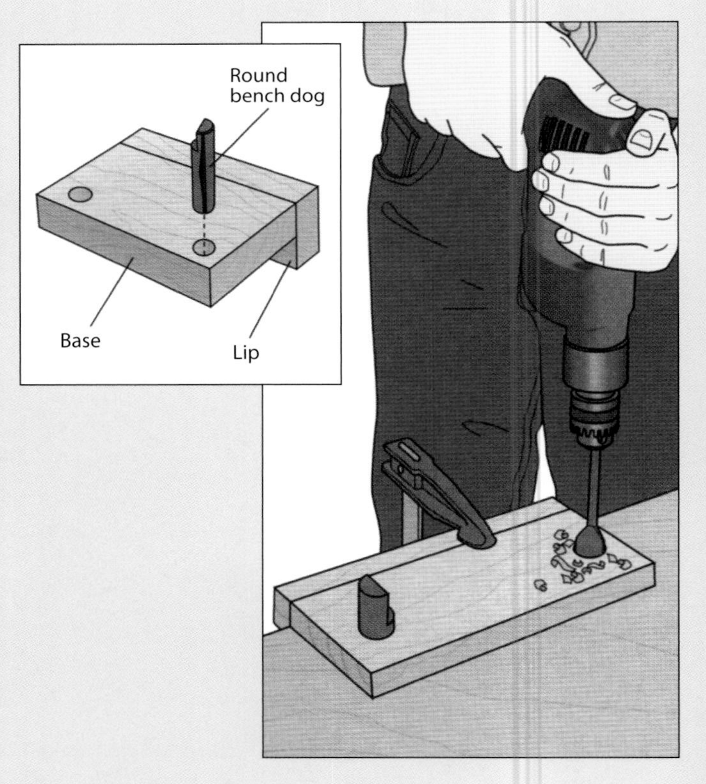

Assembling the Benchtop

Gluing up the top

First, glue up the top slab. Before gluing up the benchtop, rout grooves on both sides of the dog blocks and front rail, on one face of the front apron and back rail, and along the edges and ends of the top slab. Cut matching keys and splines. Refer to the drawing on page 33 for the size and placement of the grooves, keys, and splines. If you want to incorporate a tool tray in your bench, cut ½-inch rabbets into the bottom edges of the back rail and apron; later in the assembly process you will fit a piece of ½-inch plywood to form the tray. Set aside the sliding dog block (with the hardwood keys glued in place) and front and back aprons, spread glue on all mating surfaces, and clamp *(right)*, alternating the bar or pipe clamps on the top and bottom of the work.

Attaching the end caps

The end caps can be applied while the tail vise is being installed *(page 41)*. When that is done, invert the benchtop and rout a T-shaped recess at each end, centered between the edges. Cut two rectangular fittings from scrap hardwood so that they fit in the base of each recess. Notch one side of each fitting to accept a ⅜-inch nut, and place a fitting and nut in each recess. Set the end caps in position and mark where they contact the recesses. At each mark bore a hole for a ⅜-inch bolt, counterboring so the bolt heads are flush. Rout a groove in each end cap to accept the plywood spline, and rout a ½-inch rabbet on the bottom inside edge of the back rail to accept the tool tray. Install the tail vise on the right-hand end cap *(page 41)*. Spread glue on the contacting surfaces, fit the end caps *(left)*, and bolt them in place *(inset)*. Finally, fit the front and rear aprons and tool tray and clamp.

Vises and Accessories

Vises are the tools that transform the workbench from a simple, flat surface into a versatile work station. The modern woodworking bench incorporates two types of vise: the face vise that secures work to the front edge of the bench, and the tail vise that uses wood or metal bench dogs to secure work on the top of the bench. The pages that follow examine ways of installing both the tail vise *(page 41)* and face vise *(page 42)*.

Face vises made entirely of wood are rare. However, a wooden vise is preferable to a metal type because wooden jaws can grip work without marring its surface. A good compromise can be reached by buying the hardware for a metal vise and mounting wooden face blocks. You can extend the capacity of a face vise by boring holes in the benchtop and securing work between a bench dog in the vise's jaws and one inserted in one of the holes.

Tail vises are available in two types: an enclosed model that incorporates a sliding dog block *(below and page 41)* and one that features an L-shaped block, as in the photo at right. Some tail vises extend across the entire end of a workbench and have two screws; these are known as end vises, and they extend the utility of an already versatile tool.

Some tail vises, like the one shown above, incorporate an L-shaped shoulder block. The block allows work to be clamped between the rear jaw of the vise and the end of the bench.

Anatomy of a Tail Vise

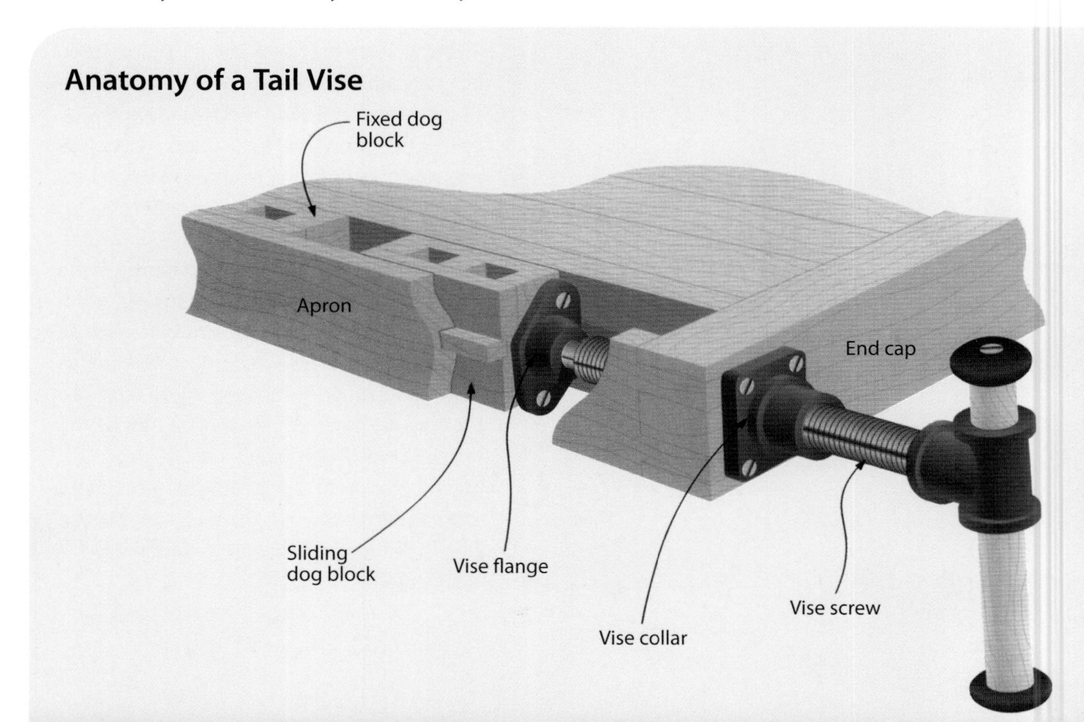

Installing a Tail Vise

Installing the vise hardware
To install a tail vise on a bench with a sliding dog block, position the vise collar against the right-hand side end cap and outline the hole for the vise screw. Then set a support board on the drill press table and clamp the end cap on top of it. Fit the drill press with a spade bit slightly larger than the vise screw and bore a hole through the end cap *(near right)*. Screw the vise collar to the end cap so the two holes line up. Next, secure the sliding dog block end-up in handscrews and clamp the handscrews to a work surface. Position the vise flange on the block and mark its screw holes. Bore a pilot hole at each mark, then screw the flange to the block *(far right)*.

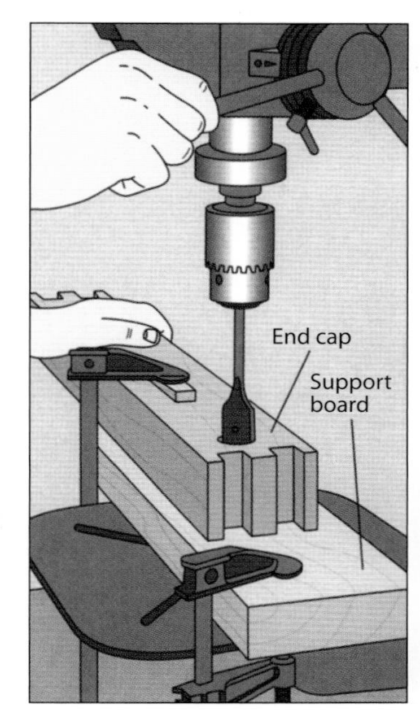

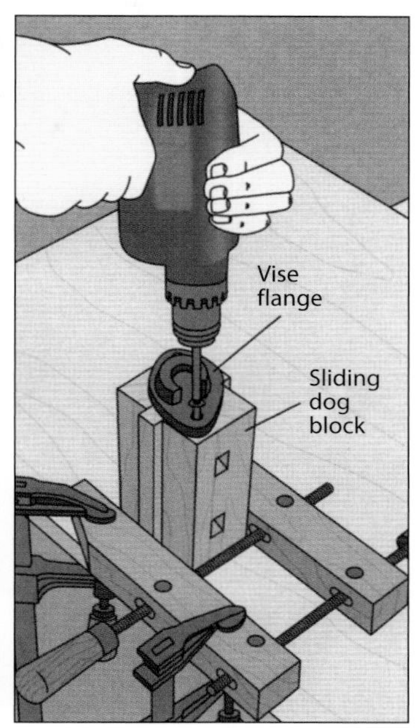

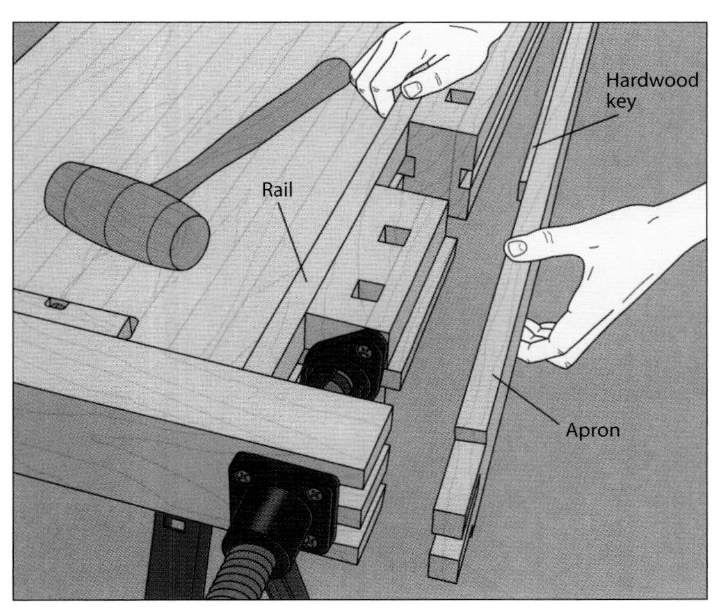

Assembling the vise
Fit the sliding dog block in the bench so the hardwood keys in the block run in the grooves in the sides of the rail. Thread the vise screw through the vise collar, test-fit the end cap on the benchtop and lock the ball joint on the end of the screw into the vise flange. Set the front apron in position against the dog blocks *(left)* and test the movement of the vise by turning the screw. If the sliding block binds, remove the end cap, apron, and sliding dog block, and ease the fit by paring the keys with a chisel. Once you are satisfied with the vise's movement, attach the aprons, end caps, and trays following the procedures outlined on page 39.

Anatomy of a Face Vise

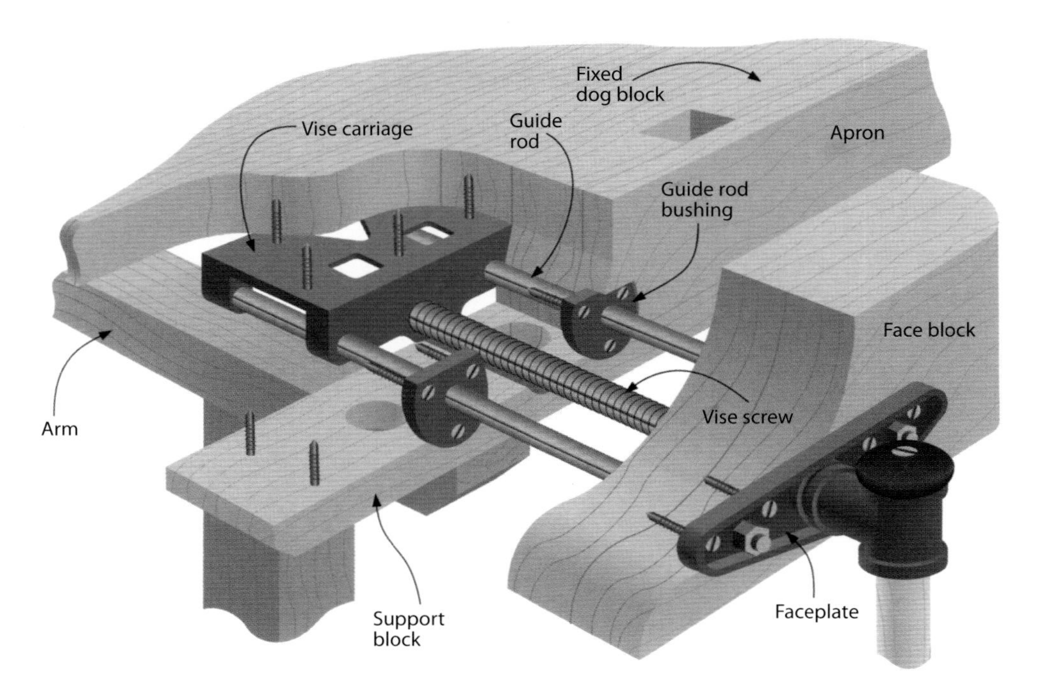

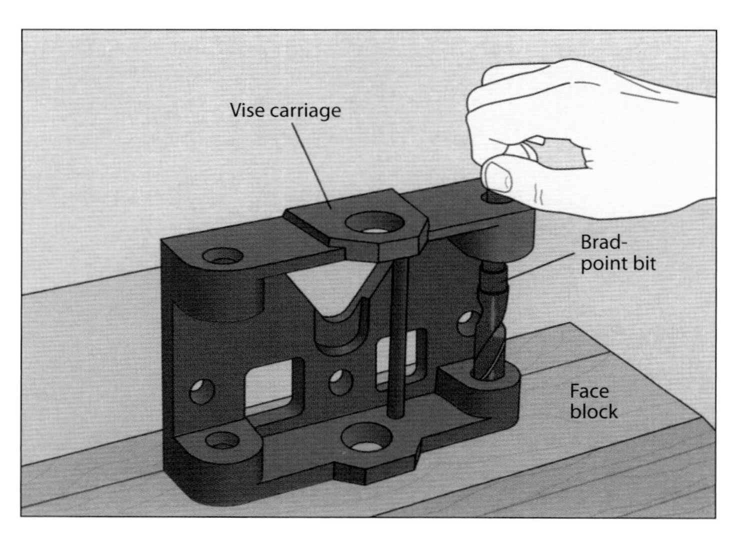

Installing a Face Vise

Preparing the face block
Cut an 18-inch-long ¾-by-3½ inch hardwood support block and screw it in place under the front left corner of the bench, after boring a row of clearance holes for the bench dogs. Next, build up the face block by gluing two pieces of 8/4 hardwood together; cut it to a final size of 5-by-18 inches. To mark and bore the holes for the vise screw and guide rods, mark a line across the face of the face block; offset the line from the top edge by the thickness of the benchtop slab (not the front apron depth). Now use the carriage as a template: Center its top edge on the line and use a brad-point bit to accurately mark the position of the three holes *(left)* and bore them.

Preparing the bench

Once the holes have been drilled through the face block, transfer their location to the workbench apron. Set the face block and benchtop on sawhorses and use bar clamps to hold the block in position against the apron; protect the stock with wood pads. Make sure the top edge of the block is flush with the benchtop and its end is flush with the end cap. Mark the hole locations on the apron using the brad-point bit *(right)*. Remove the face block and bore the holes through the apron and bench dog block.

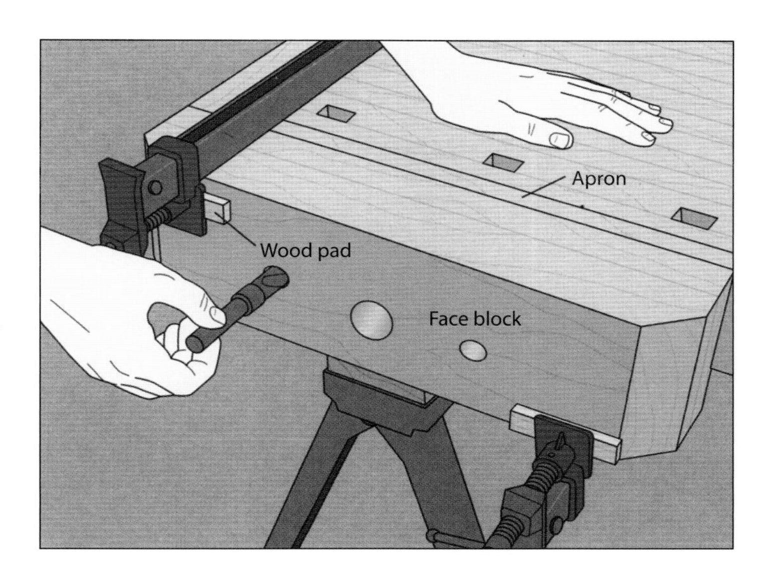

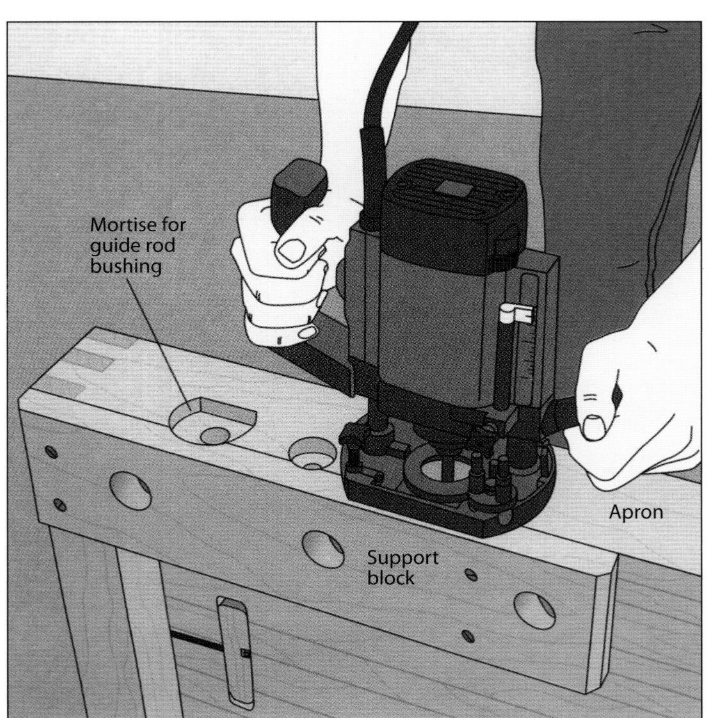

Mounting the vise

Attach the vise assembly—the face-plate, screw, and guide rods—to the face block. Turn the benchtop upside down, place the vise carriage on the bench's underside, and feed the vise screw and guide rods through the holes in the apron and into the carriage. Make pilot holes on the underside of the bench and fasten the carriage in place. Next, fasten the guide rod bushings to the apron: Remove the vise assembly, fit the bushings on the rods, remount the assembly, and outline the bushings' location on the apron. Then remove the vise assembly again and secure the benchtop so the apron is facing up. With a router and straight bit, cut recesses for the bushings within the outlines *(left)*. Screw the bushings to the recesses in the apron and attach the vise to the bench. Now the workbench top is ready to be attached to the base. Lay the top upside down on the floor, place the base in position, and drive lag screws through the arms into the top.

Jigs for Iron-Jawed Bench Vises

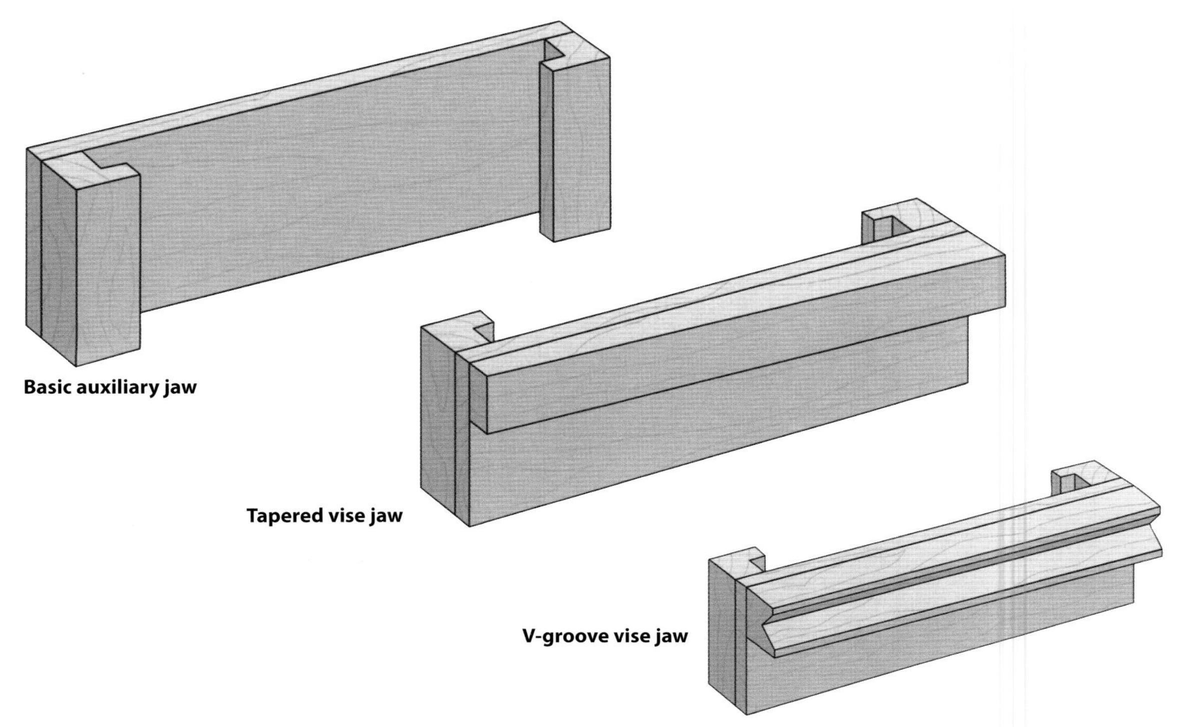

Fitting wooden inserts to metal jaws

If your bench is equipped with a metal-jawed vise like the one shown at the top of page 45, fitting interchangeable auxiliary jaws can extend the vise's versatility. The wooden inserts shown above will not only be less damaging to workpieces than metal jaws, but they can also be custom-made for special jobs. Each insert is made from ½-inch-thick solid stock with a rabbeted 1-by-1 block glued at each end to hug the ends of the vise jaw. Although a pair is required, only one of each sample is illustrated. The basic jaw *(above, left)* will do most standard clamping jobs. The tapered jaw *(above, center)* features a wedge-shaped strip for holding tapered stock efficiently. The V-groove jaw *(above, right)* includes a strip with a groove cut down its middle for securing cylindrical work.

Shop Tip

A quick-switch vise

If you are reluctant to bolt your bench vise onto your workbench, attach it instead to a T-shaped base made of ¾-inch plywood. Join the two pieces of the base together with a dado joint and screws. Secure the vertical part of the base in either the tail or face vise of the bench.

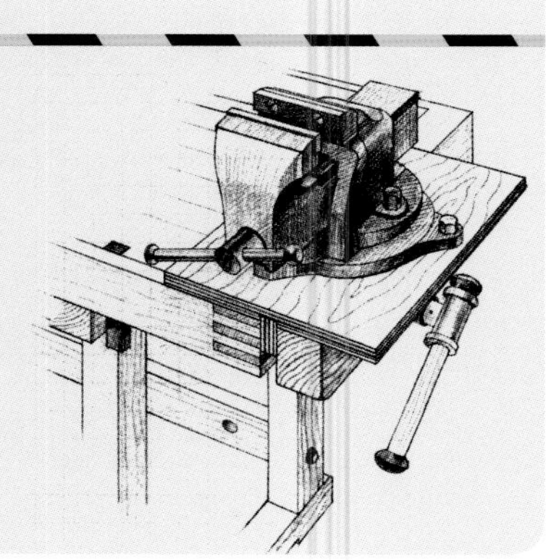

A Sliding Bench Stop

If your "workbench" is a standard table with a bench vise fastened to one edge, the jig and fence shown at right can lend it some versatility. Cut the auxiliary vise jaws from 1-inch stock and the pieces of the T-shaped vise jig from ¾-inch wood. You will need two pieces for the jig: a top and a lip. Rout a dado across one auxiliary jaw to accommodate the lip of the vise jig and another on the underside of the jig top. Screw the auxiliary jaws to the vise jaws, making a cut for the vise screw if necessary, then glue and screw the lip to the top of the jig. Cut the sliding fence from ½-inch-thick stock and cut two stopped grooves through it for ¼-inch carriage bolts. To mount the fence, bore two holes through the table for the bolts, feed the bolts through the holes and the grooves and fasten them with washers and wing nuts. To use the jig and fence, slide the lip into the auxiliary jaw, adjust the sliding fence to hold the workpiece snugly and clamp it in place by tightening the vise jaw and wing nuts.

Top
Sliding fence
Lip
Auxiliary vise jaw

Preventing Vise Racking

Using a stepped block

When securing a workpiece at one end of a face vise, the other end of the vise is likely to rack—or tilt toward the bench—and cause the work to slip. To prevent racking, use a stepped hardwood block to keep the jaws square. Cut a series of steps in one face of the block, spacing them at equal intervals, such as ½ inch. Place the block in the open end of the vise at the same time you are securing the workpiece so that the vise is parallel to the edge of the bench *(right)*.

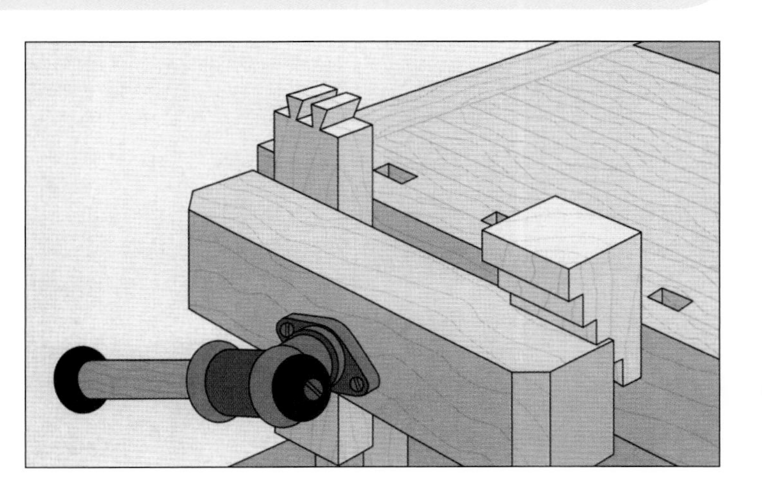

Bench Dogs and Hold Downs

Bench dogs are as important as vises in maximizing the flexibility and utility of a well-designed workbench. A set of bench dogs works like a second pair of hands to secure workpieces for planing, chiseling, mortising, carving, or other woodworking tasks.

Although the bench dog looks like a deceptively simple peg, it incorporates design features that enable it to hold a workpiece firmly without slipping in its hole. One feature usually is a thin metal spring attached to one side that presses against the inside wall of the dog hole in the workbench. To help strengthen the grip of bench dogs, the holes are also angled toward the vise at 4°.

Bench dogs can be either round or square. Round dogs are easier to incorporate in a bench that does not yet have dog holes; it is simpler to bore holes than to make square dog holes. Since round dogs can swivel, their notched, flat heads enable them to clamp stock in practically any direction. This can be a disadvantage: Some woodworkers claim that round dogs tend to slip in their holes more than square dogs, which cannot rotate.

Bench dogs can be made of either metal or wood. Metal dogs have a weight, strength, and stiffness that wooden ones cannot match. Yet wooden dogs have their advantages—as any woodworker who has nicked a plane blade on a metal dog will attest.

This bench dog features a threaded screw that converts it into a miniature tail vise. Used in conjunction with other bench dogs, it excels at clamping small or irregular work, like the panel shown above

Bench dogs are not the only method of securing stock; bench hooks, carving hooks, wedges, and hold downs are also useful for keeping stock in place. The following pages illustrate a number of commercial and shop-made options to keep workpieces put while you work.

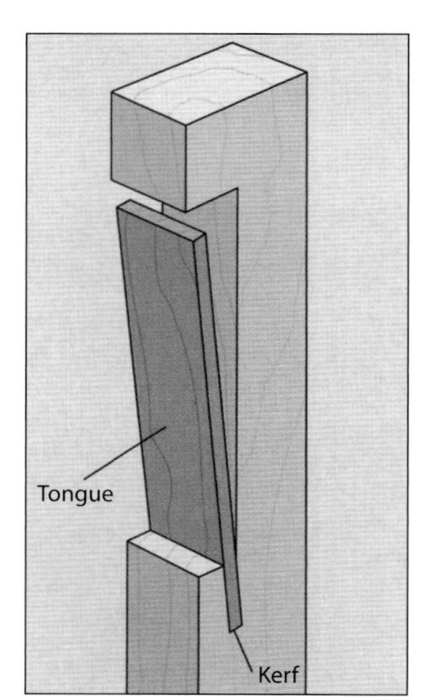

Bench Dogs

Making a wooden bench dog

Bench dogs can be crafted from hardwood stock; the one shown at left uses an angled wooden tongue as a spring. Cut the dog to fit the holes in your workbench, then chisel out a dado from the middle of the dog. Saw a short kerf into the lower corner of the dado, angling the cut so the tongue will extend beyond the edge of the dado. Cut the tongue from hardwood, making it about as long as the dado, as wide as the dog, and as thick as the kerf. Glue the tongue in the kerf.

Making a spring-loaded bench dog

A wooden bench dog can be made to fit snugly by equipping it with a metal spring cut from an old band saw or hacksaw blade. Cut your dog to size, then chisel out a small recess for the spring. The width and depth of the recess should equal the width and thickness of the spring, but its length should be slightly shorter than that of the spring. Press the spring into the recess; the metal will bow outward, holding the dog firmly in its hole.

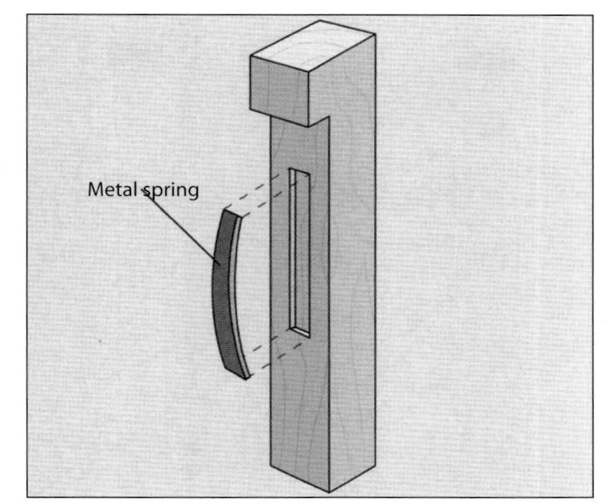

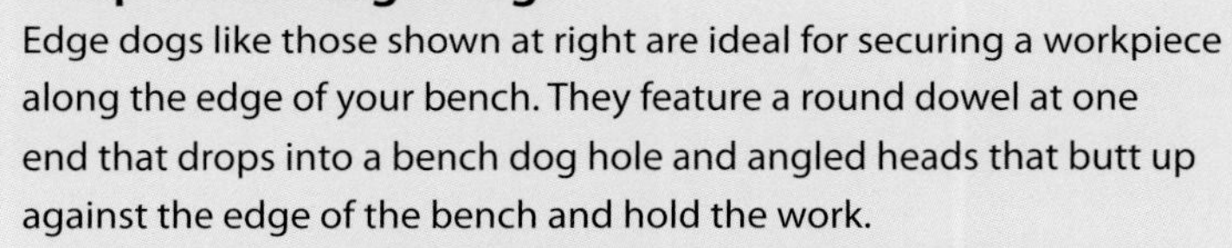

Shop-Made Edge Dogs

Edge dogs like those shown at right are ideal for securing a workpiece along the edge of your bench. They feature a round dowel at one end that drops into a bench dog hole and angled heads that butt up against the edge of the bench and hold the work.

Start by cutting the dogs from hardwood stock. Both left-hand and right-hand dogs are needed, with the heads angled in opposing directions. Bore a ½-inch-diameter hole through the ends, and drive a 3-inch length of dowel in each hole. Then insert the dowel in a bench dog hole and angle the dog so it extends beyond the edge of the table. Mark a 90° notch for the head perpendicular to the edge of the bench and cut it out. To hold the edge dog in place when clamping pressure is applied, saw a ¼-inch-slice off the bottom of the dog, except for the head. This provides a lip that will butt against the edge of the bench *(top left)*.

To use the edge dogs, place the left-hand dog in a hole in the fixed dog block and the right-hand one in the sliding dog block of the tail vise. Tighten the vise until the workpiece is held in the notches *(bottom left)*.

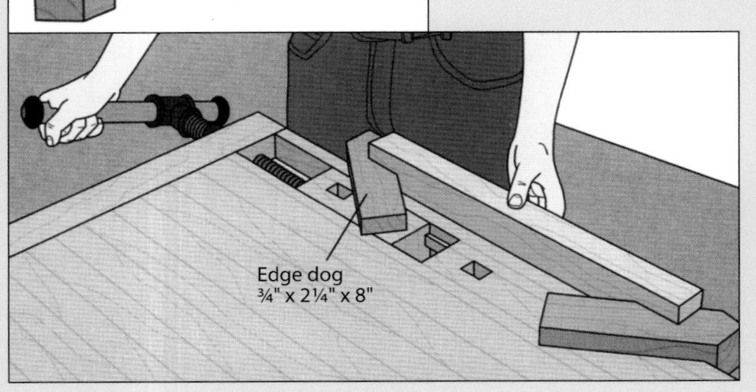

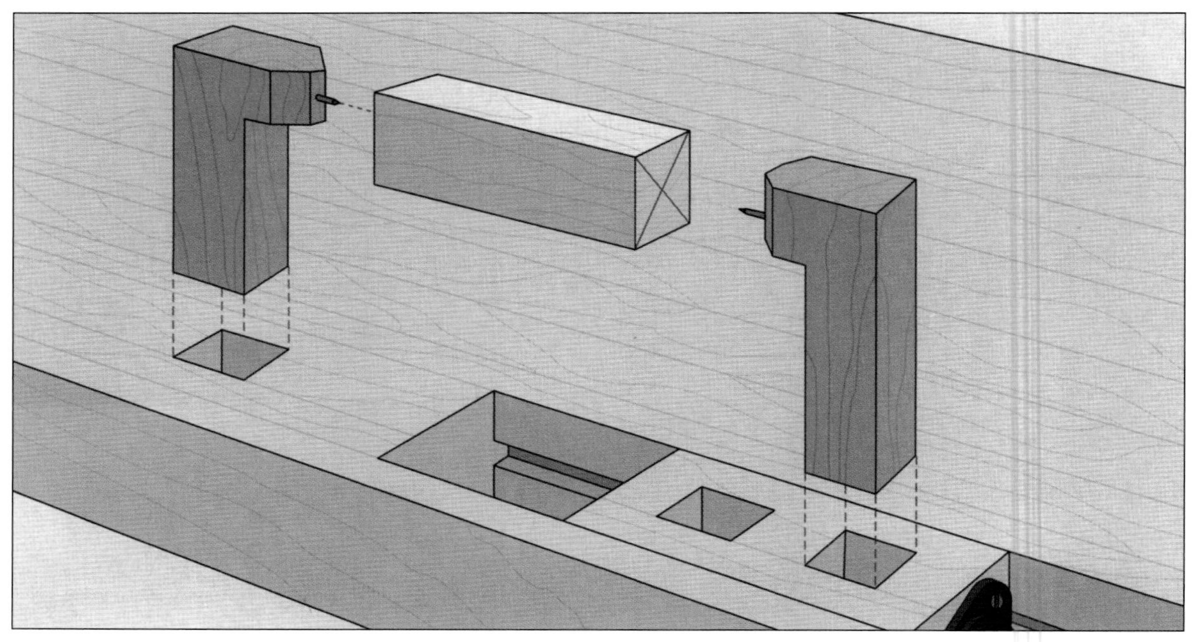

Making and setting up carving dogs

Using a standard bench dog as a model, you can fashion a pair of customized dogs that will grip a carved or turned workpiece, or secure irregular-sized work, such as mitered molding. To make these accessories, cut bevels on either side of the head of a standard bench dog and drive a small screw or nail into the center of the head; snip off the fastener's head to form a sharp point. To use the devices, place one dog in a dog hole of the bench's fixed dog block and the other in the tail vise or a sliding dog block hole *(above)*. Tighten the vise screw until the points contact the ends of the workpiece and hold it securely.

Shop Tip

Carving screws

A pair of hanger bolts can enable you to secure an irregular-shaped workpiece, such as a carving block, to your bench. The bolts feature wood screw threads on one end and machine screw threads on the other. To secure a workpiece, bore two holes through the benchtop for the bolts. Screw the bolts into the carving block from underneath the top and hold the bolts to the underside of the top with washers and wing nuts.

Hold-Downs and Bench Stops

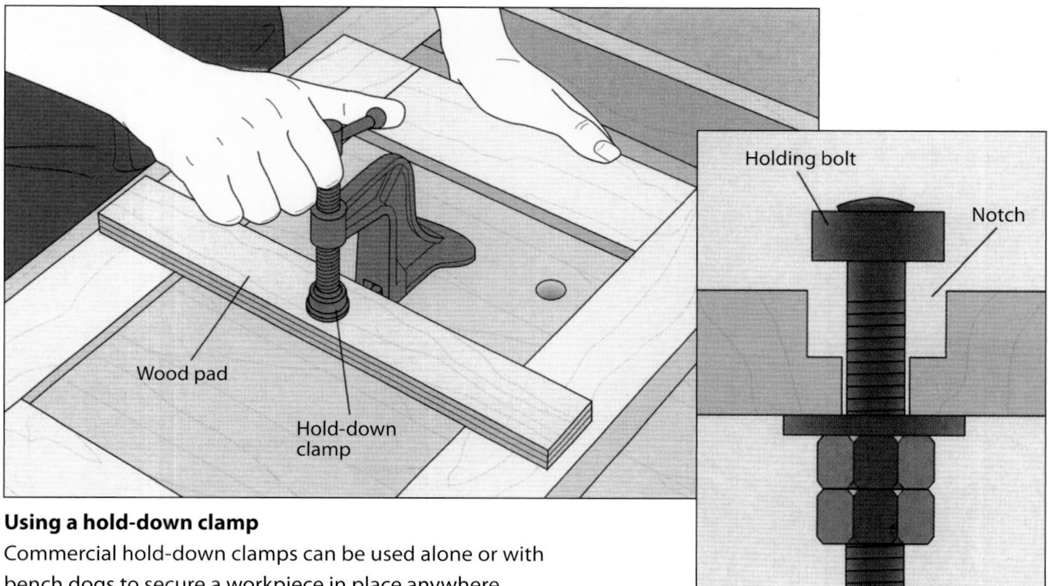

Using a hold-down clamp

Commercial hold-down clamps can be used alone or with bench dogs to secure a workpiece in place anywhere on a workbench. The type shown features an adjustable holding bolt which sits in a counterbored hole through the benchtop *(inset)*. To use the clamp, raise the bolt head and slide it through the notch at the base of the clamp. Set the workpiece under the clamp jaw and tighten the screw *(above)*. (In the illustration, a wood pad is being used to apply equal pressure to both stiles of a door frame.) To remove the clamp from the bench, slide it off the bolt head and let the bolt drop below the surface of the top.

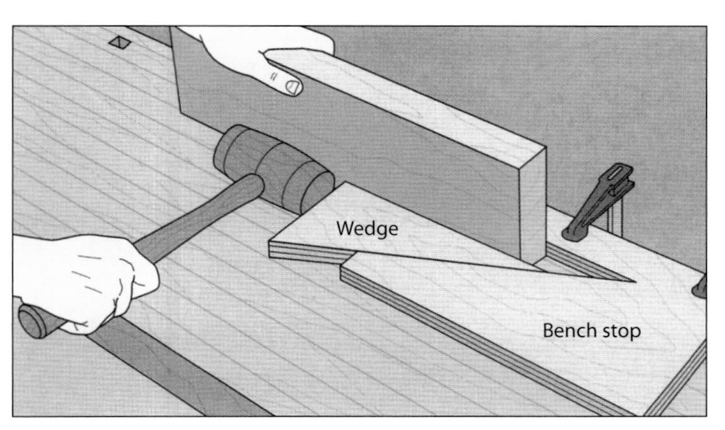

Making and using a temporary bench stop

A clamped-on bench stop cut from ¾-inch plywood will secure a workpiece to the benchtop without the help of bench dogs. Cut the bench stop to size, then mark out a triangular wedge, typically 3 inches shorter than the stop. Cut out the wedge and set it aside. To use the bench stop, clamp it to the benchtop and slide the workpiece into the notch, butting one side against the straight edge of the notch. Secure the piece with the wedge, tapping it tightly in place with a mallet *(left)*.

Installing a wedge stop

A wedge stop can also be used to secure stock on a benchtop *(right)*. The stop consists of a fixed rail and a movable rail that are secured by dowels resting in a double row of holes bored into the workbench. Together with a triangular wedge, the rails keep a workpiece from moving. Cut the rails and the wedge from ¾-inch plywood. (You can choose thicker stock for the rails, depending on the thickness of your workpiece.) Bore two ½-inch-diameter holes in each rail, then glue a 2-inch-long dowel in each hole. Bore two rows of ½-inch-diameter holes in the workbench for the dowels. To use the stop, place the fixed rail at one end of the row of holes and the movable rail the appropriate distance away so the wedge, when positioned between the rails, will keep the workpiece steady.

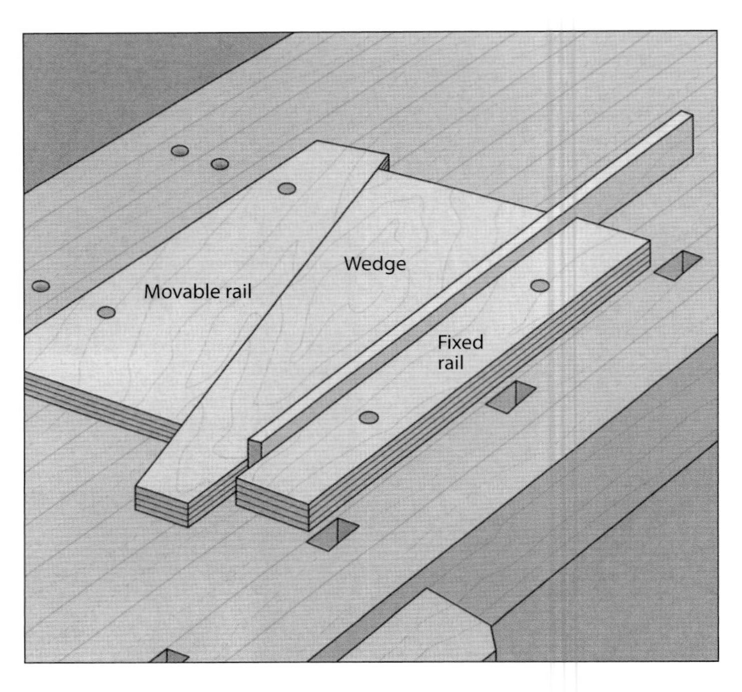

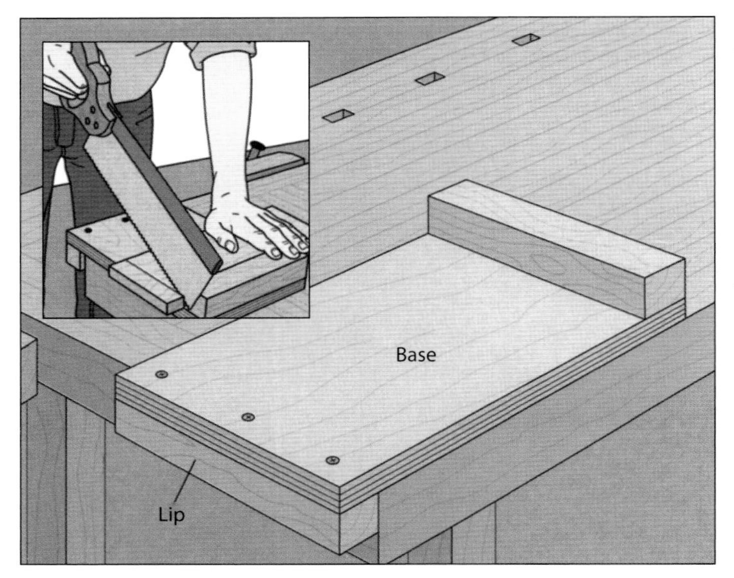

Making a bench hook

The shop-built jig shown at right will ensure that the crosscuts you make on the workbench will be square. Use ¾-inch plywood for the base and strips of 2-by-2 stock for the lips. Make the base at least as long as the width of your workpiece and wide enough to support it. Screw the lips to the guide, attaching one to each face. To use the jig, butt one lip against the edge of the bench and press the workpiece firmly against the other. Align the cutting line with the edge of the base and make the cut *(inset)*.

Making a flip-up stop

The flip-up bench stop shown at right provides another way to make quick guided crosscuts on a workbench. Cut the two pieces of the stop from hardwood. Screw the pieces to the end of the benchtop; on the bench shown, the inner edge of the pivoting piece is lined up with the edge of the tool tray to provide a convenient reference line for squaring up a crosscut. Screw the stationary piece in place with two screws, and the flip-up piece with one so that it can pivot. When not in use, the pivoting piece should lie on edge atop the stationary piece. To use the stop, flip up the pivoting piece, butt the workpiece against it, and make your crosscut.

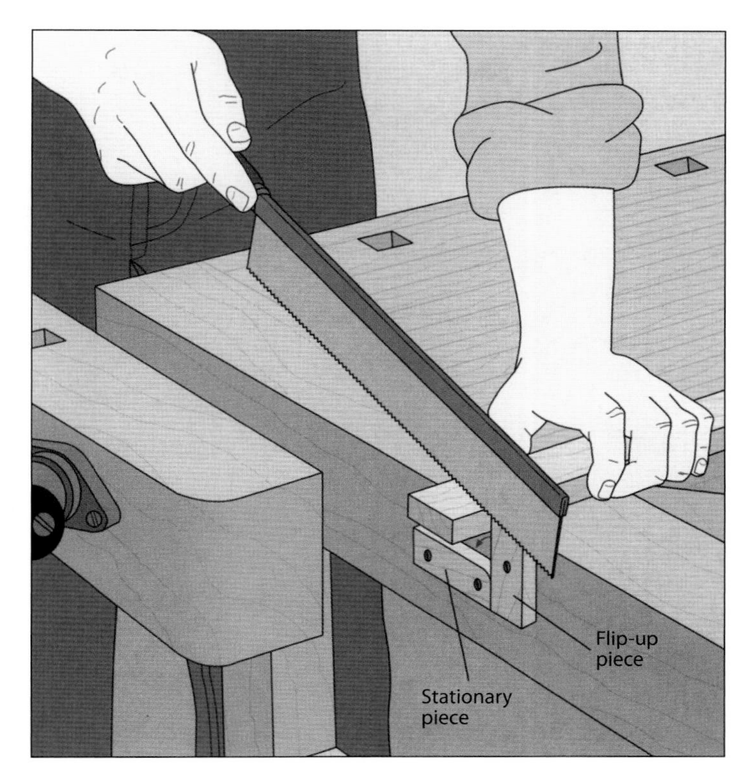

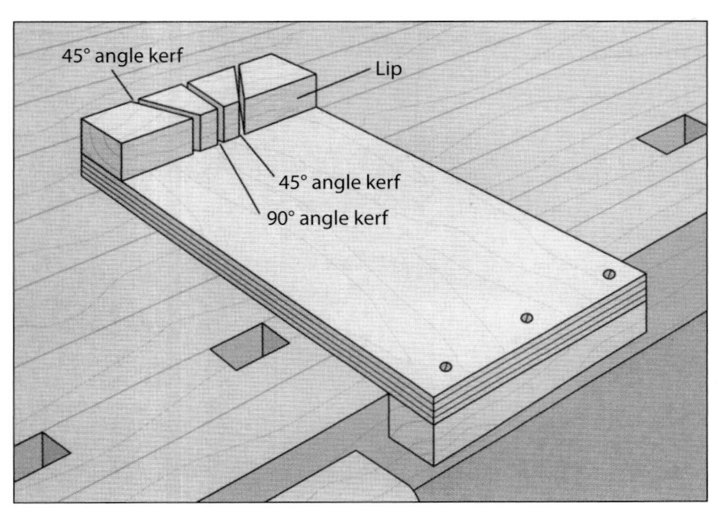

Making and using a miter bench hook

Customize a standard bench hook to make 45° angle miter cuts by adding kerfs to one of the lips. Build a bench hook *(page 50)*, then use a backsaw to cut two kerfs in the lip at opposing 45° angles and one at 90° *(left)*. Use the miter bench hook as you would a standard bench hook, lining up the cutting line on the workpiece with the desired kerf.

A Substitute Bench Vise

If your workbench does not have a bench vise, you can improvise a substitute using readily available shop accessories. Two large handscrews arranged as shown above will hold a board upright at the corner of the work surface.

Shop Tip

Gripping thin stock
Securing a thin workpiece on edge usually requires a vise or bench dogs. However, you can fashion a bench stop like the one shown on page 90 to accomplish the task. In this case, make the stop from thicker stock—about 2 inches thick—to get a better grip on the workpiece and locate the wedge closer to the middle of the stop. Clamp the jig to the benchtop.

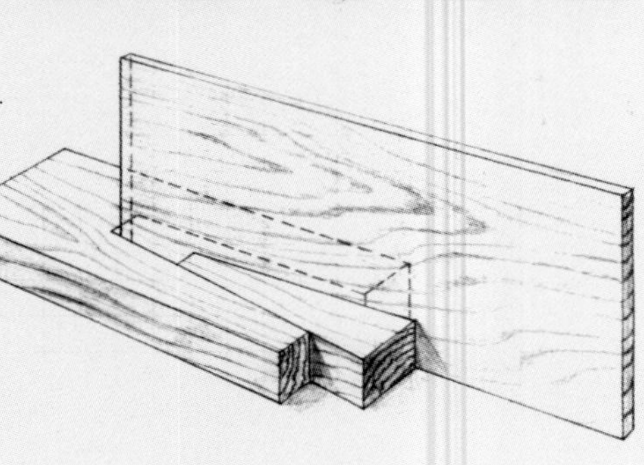

Two Shooting Boards

Making shooting boards
To smooth end grain with a plane, use a shooting board like those shown at left. The right-angle shooting board *(left, above)* is for planing straight end grain; a mitered version can also be built *(left, below)*. Cut the pieces according to the dimensions suggested in the illustrations. Build the base, top, and mitered stop block from ¾-inch plywood; use solid wood for the lip and the square stop block. Screw the top to the base with the ends and one edge aligned. Then attach the lip to the base, making sure that the lip lines up with the end of the base. For the right-angle shooting board, fasten the stop block to the top flush with the other end of the jig. For the mitered shooting board, center the stop block on the top.

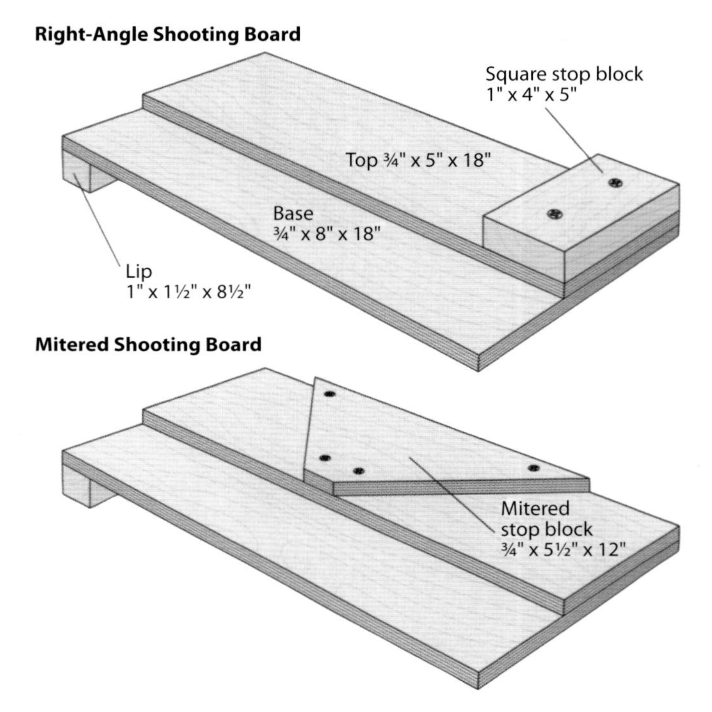

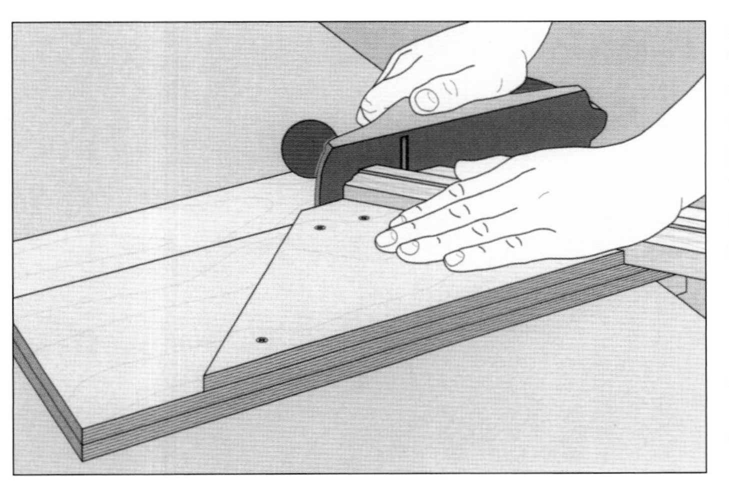

Smoothing end grain
To use either jig, hook the lip on the edge of a work surface. Set your workpiece on the top, butting the edge against the stop block so that it extends over the edge of the top by about 1⁄16 inch. With the mitered shooting board, position the workpiece against the appropriate side of the stop block. (For a long workpiece, it may be necessary to place a support board under the opposite end to keep the workpiece level.) Set a plane on its side at one end of the jig and butt the sole against the edge of the top. Holding the workpiece firmly, guide the plane along the jig from one end to the other *(left)*.

Shop Accessories

Look beneath the surface of an efficient, well-equipped shop, and you will find several invisible auxiliaries: accessories designed to make the work safer and the shop more comfortable in which to work. The most commonly found helpers are compressors, generators, bench grinders, and—perhaps the most important for safety and comfort—dust collection systems.

Air compressors first were utilized by woodworkers only for finishing work—to apply lacquer and varnish more smoothly than with a brush. But with the advent of such tools as pneumatic nailers, compressors are found more frequently, even in small home workshops. Air-powered tools are discussed starting on page 58.

Grinders, of course, can speed tool sharpening. More importantly, as you will see on page 60, they can permit you to modify tools and reclaim damaged cutting edges.

Airborne dust once was considered an unavoidable consequence of working with wood. But the increased emphasis on environmental health has led to the introduction of efficient dust collection systems that are affordable to home woodworkers. They should be a high priority item for every home workshop. Tiny wood dust particles can remain in the shop for more than an hour after the tool has been used. The dust poses several health risks. If the wood contains toxins or irritants—and many species do—the effects can lead to a wide range of ailments, including dermatitis, shortness of breath, and dizziness. Recent studies have shown that long-term inhalation of wood dust is at least a contributing factor in cancers of the tongue, tonsils, lung, and larynx.

When you add to the equation the fire risk and the hazard of a dust-covered shop floor, there are compelling reasons for installing some kind of dust collection system in your shop. Pages 62 to 71 provide you with information you will need to set up and maintain both central and portable systems. Remember that designing a central system requires careful attention to detail and precise calculation of your specific requirements. To be safe, check your plans and figures with an engineer before installing the system.

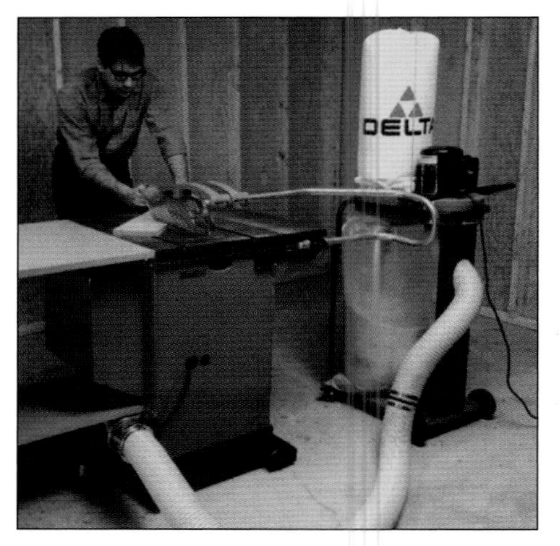

Most of the wood chips and sawdust generated by this 10-inch table saw are captured by a portable dust collection system. Often neglected in the past, dust collection has become a central concern of many safety-conscious woodworkers in planning the layout of their shops.

Hooked up to a compressor, this air-powered sander is compact enough to hold in one hand, yet it smooths wood as efficiently as an electric sander.

A Store of Shop Accessories

Planers can create a substantial mound of sawdust in short order. A portable dust collector will keep most of the dust from this and other power tools off the shop floor and out of the air.

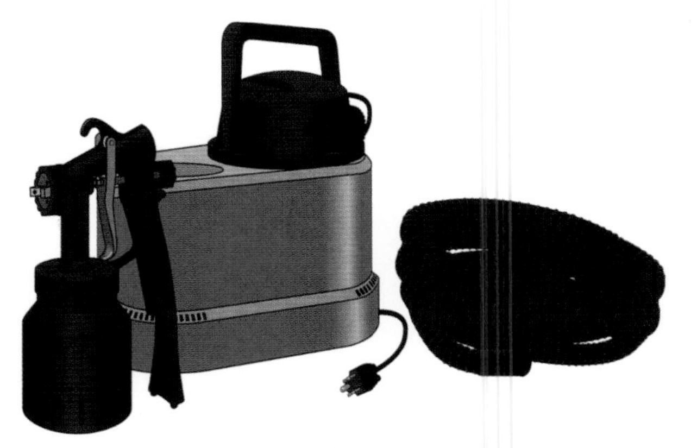

High-volume, low-pressure (HVLP) spray system
For applying stains and finishes. Features electric turbine that supplies large amount of air at low pressure through air hose to spray gun; compared to conventional, compressed-air type systems, HVLP allows higher percentage of finish to contact workpiece.

Air compressor
Supplies stream of high pressure air through hose to power a variety of air-operated tools, such as sanders, spray guns, and drills; consumer-grade models range from ⅛ to 5 horsepower and can generate up to 200 pounds per square inch (psi) of air pressure and 0.3 to 15 cubic feet of air per minute (cfm).

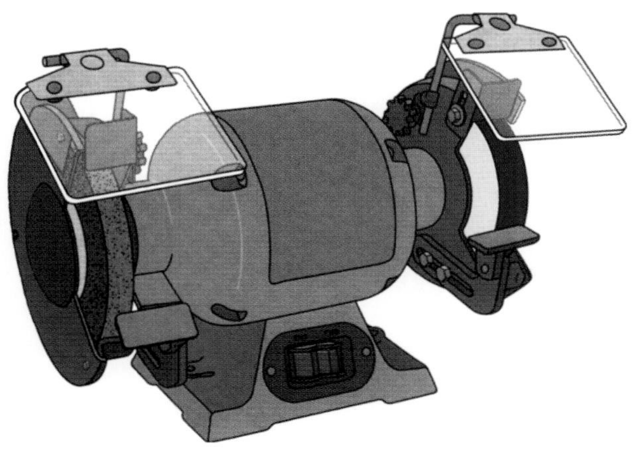

Bench grinder
Coarse wheel *(left)* squares, sharpens, and smooths blades and bits; cloth wheel *(right)* polishes and cleans. Features a ¼- to ½-horsepower electric motor; eye shields, adjustable tool rests, and wheel guards standard on most models. Benchtop grinders usually bolted to work surface.

Shop vacuum
Cleans up dust and liquid spills; hose can be attached to individual tools to collect dust as it is produced. Typically features 1¼- to 2½-inch-diameter collection hose and 5- to 10-gallon tank; some models can double as portable blower.

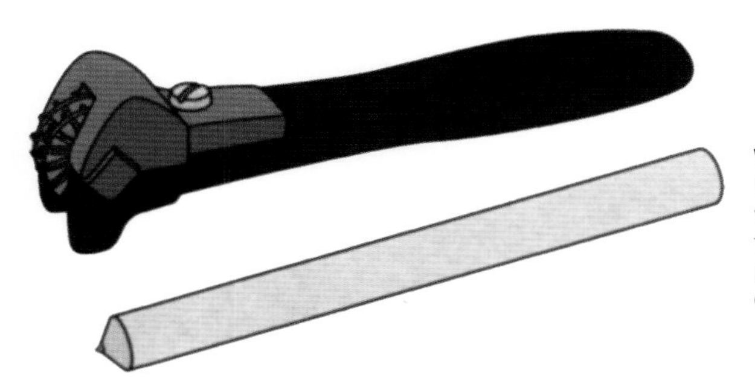

Wheel dresser
Used to true or reshape bench grinder wheel. Star-wheel dresser *(top)* uses up to four star-shaped wheels; diamond-point dresser *(bottom)* features ¼-carat diamond set in bronze tip and metal shaft.

Air Compressors

An air compressor can be fitted with a large number of tools and attachments, making it a convenient shop accessory. In some shops, a compressor can represent an alternative to some electric tools. For others, it can be a valuable supplement.

Pneumatic drills, grinders, sanders, and wrenches perform at least as effectively as their electric-powered counterparts. Some tools, like sprayers, nailers, and abrasive cleaners, are clearly superior to the alternatives.

Compressors and the tools they drive are inherently simple: The air is drawn in, pressurized by a diaphragm or one or more pistons, and usually stored in a tank. When the trigger on an air-driven tool is pressed, the air travels through a hose to power the tool.

Because they contain no heavy electric motor, most air tools are lighter, cheaper, and easier to repair than their cousins. They cannot overheat, and there is no danger of electrical shock.

Compressed-air power does have some drawbacks, chiefly the cost of the compressor itself and maintenance. Air drills and the like must be oiled daily. And you will invest several hundred dollars in a compressor that is capable of driving typical shop tools.

Some air-powered tools require a sizable volume of air, usually measured in cubic feet per minute (cfm); others need a minimum level of air pressure in pounds per square inch (psi). When choosing a compressor, consider the cfm or psi requirements of the air-powered tools you plan to use and buy a compressor with slightly more power. You never know when you will want to expand your tool inventory.

Air-Powered Tools and Accessories

Jitterbug sander
Orbital sander capable of producing 2500 strokes per minute; weighs less than 5 pounds. Requires 6.5 cfm at 90 psi; must be used with tank-mounted compressor with at least 3 horsepower.

Drill
⅜-inch drill that turns bits at 2500 rpm; weighs only 2¼ pounds. Requires 3 cfm at 90 psi; must be used with tank-mounted compressor with at least ¾ horsepower.

Spray gun
Heavy-duty sprayer with adjustable fluid and air controls. Requires 5.5 cfm at 40 psi; can be used with any compressor with more than 1 horsepower.

Compressor Safety Tips

- Read your owner's manual carefully before operating a compressor or any air-powered tool.
- Do not reset any switches or valves on the compressor; they have been preset at safe levels at the factory.
- Check the hoses, plugs, wires, pipes, and tubes of the compressor, and the tool air inlets before each use. Do not use the compressor or tool if any part is worn or damaged.
- Wear safety glasses and hearing protection when using air-powered tools.
- Do not exceed the pressure rating of an air tool or accessory.
- Always plug a compressor into a grounded outlet of the appropriate amperage.
- Relieve pressure slowly when depressurizing the tank.
- Do not press the trigger of an air tool when connecting it to an air hose.
- Do not remove the belt guard of a belt-driven compressor when the machine is operating.
- Turn the compressor off if it produces an unfamiliar noise or vibration, produces insufficient air pressure, or consumes excessive oil; have the machine serviced before resuming operations.
- Allow the compressor to cool before performing any maintenance; wear gloves to disconnect any parts that are still hot.
- Turn the compressor off before moving it.
- Do not touch the compressor while using it or immediately after; the machine can become very hot.
- Drain any moisture from the tank after each use to prevent rust; tank pressure should be no higher than 10 psi when draining it.
- Replace the tank if it has any pin holes, rust spots, or weak spots at welds.

Brad finishing nailer
Nail gun for driving ⅜- to 1¼-inch No. 18 finishing nails; weighs less than 3 pounds. Narrow nose sets nails without marring workpiece; magazine holds up to 110 nails. Requires .28 cfm at 90 psi to drive 10 nails per minute; must be used with compressor with at least ½ horsepower.

Hose connector
Joins air tool to compressor or connects two lengths of compressor hose together.

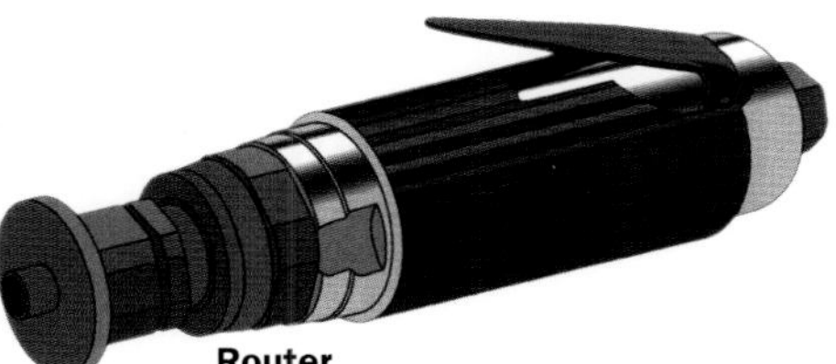

Router
¼-inch direct-drive router that turns bits at 20,000 rpm; weighs just over 1 pound. Features neoprene rubber grip to reduce vibration. Requires 90 psi; will function with most compressors.

Quick coupler
Used with hose connectors to attach air tools to compressor hose or to join lengths of compressor hose together; automatically shuts off air when uncoupled from compressor.

Bench Grinders

From dressing and shaping metal to squaring and sharpening bits, plane irons, and chisel blades, the bench grinder is an invaluable workshop maintenance tool. Grinders are classified according to their wheel diameter. The 6-to 8-inch benchtop models, with ¼- to ½-horsepower motors, are the most popular home workshop sizes. They can be mounted on a work surface or fastened to a separate stand.

Grinding wheels come in many grits and compositions. Medium 36- and 60-grit aluminum oxide wheels will handle most tasks adequately, but you may need a finer wheel, with either 100 or 120 grit, for delicate sharpening jobs. Buffing wheels for polishing metal, and wire wheels for removing rust and cleaning metal, are also worth owning.

Most grinders operate at one speed, or allow a choice of two speeds—typically 2950 and 3600 rpm.

No grinder should be used without lowering the guard mounted above each wheel; the tool should also come equipped with adjustable tool rests and wheel covers sheathing 75 percent to 80 percent of the wheels.

A grinder is the best tool for restoring the correct bevel angle on a nicked or out-of-square chisel blade. The tip of the blade must contact the grinder wheel at an angle of 25° to 30°.

Dressing a Grinder Wheel

Truing the wheel

To true a grinder wheel and square its edges, use a star-wheel dresser or a diamond-point dresser. For the star-wheel dresser, move the grinder's tool rest away from the wheel. With the guard in position, switch on the grinder and butt the tip of the dresser against the wheel. Then, with your index finger resting against the tool rest, move the dresser side-to-side across the wheel *(left)*. For the diamond-point dresser, hold the device between the index finger and thumb of one hand, set it on the tool rest, and advance it toward the wheel until your index finger contacts the tool rest *(inset)*. Slide the tip of the dresser across the wheel, pressing lightly while keeping your finger on the tool rest. For either dresser, continue until the edges of the wheel are square and you have exposed fresh abrasive.

Gouge-Sharpening Jig

The jig shown at right guarantees that the tip of a gouge will contact the wheel of your grinder at the correct angle to restore the bevel on the cutting edge. The dimensions in the illustration will accommodate most gouges.

Cut the base and the guide from ½-inch plywood. Screw the guide together and fasten it to the base with screws countersunk from underneath. Make sure the opening created by the guide is large enough to allow the arm to slide through freely.

Cut the arm from 1-by-2 stock and the tool support from ½-inch plywood. Screw the two parts of the tool support together, then fasten the bottom to the arm, flush with one end. For the V block, cut a small wood block to size and saw a 90° wedge out of one side. Glue the block to the tool support.

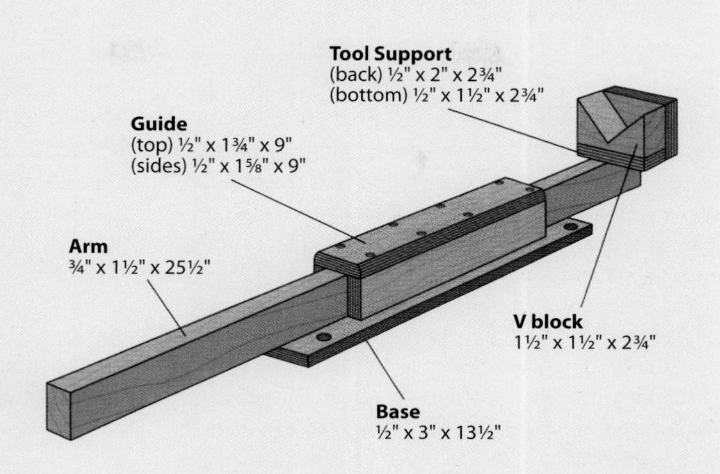

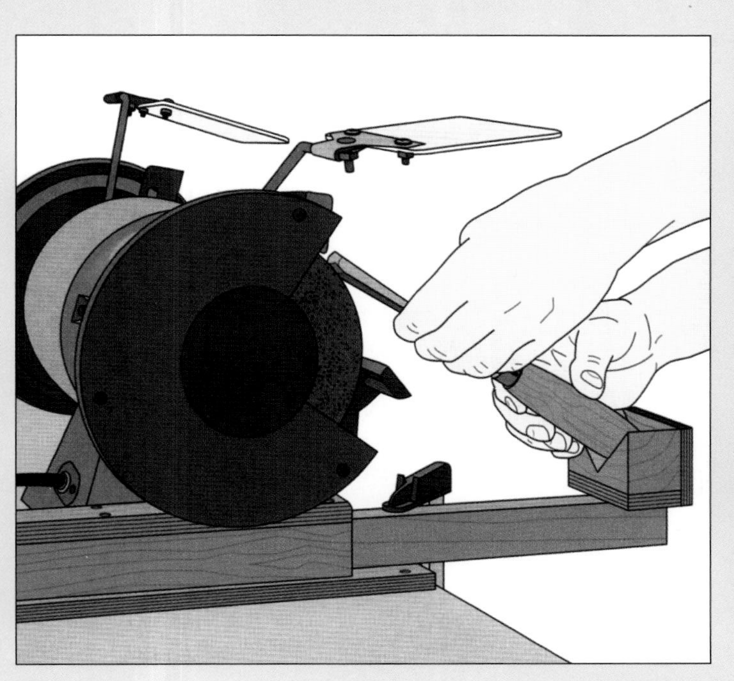

To use the jig, secure it to a work surface so the arm lines up directly under the grinding wheel. Seat the gouge handle in the V block and slide the arm so the beveled edge of the gouge sits flat on the grinding wheel. Clamp the arm in place. Then, with the gouge clear of the wheel, switch on the grinder and reposition the tool in the jig. Holding the gouge with both hands, rotate it from side-to-side so the beveled edge runs across the wheel *(left)*. Check the cutting edge periodically and stop grinding when the bevel forms.

Dust Collection

A dust collection system has one aim: to capture most of the wood dust created at each of your woodworking machines and prevent it from ending up on the shop floor, or, worse yet, in the air. There are a series of variables in every system that must be coordinated to ensure a strong enough flow of air: the power of the collector; the location and requirements of the machines in the shop; and the type, size, and layout of the duct work.

The design of a central system begins with a simple bird's-eye view sketch of your shop, like the one shown below, arranging the machines and collector in their preferred locations. Then, draw in a main line running from the collector through the shop. Sketch in branch lines as needed to accommodate each machine and any obstructions—joists, beams, or fixtures—that may require special routing. For the best air flow, keep the main line and branch lines as short and straight as possible, and position the machines that produce the most dust closest to the collector. You may choose to run ducting along the ceiling of the shop, or, to increase the efficiency of the system, at machine-table height along the walls.

Since in most home shops only one woodworking machine will be producing dust at a time, 4- or 5-inch-diameter duct is sufficient for both the main and branch lines. There are several suitable types of duct available for dust collection systems. The best choice is metal duct designed specifically for dust collection. However, many woodworkers opt for plastic pipe, typically PVC or ABS. It is easier to seal and assemble (and disassemble for cleaning), less expensive, and more readily available.

Because plastic is an insulator, however, static build-up inside the pipe can reach dangerous levels during use—possibly high enough to ignite the dust passing through it. To prevent this, ground all plastic ducts by running a bare copper ground wire from each tool, inside the duct, to an electrical ground. As a safety precaution, have the system checked by an electrician.

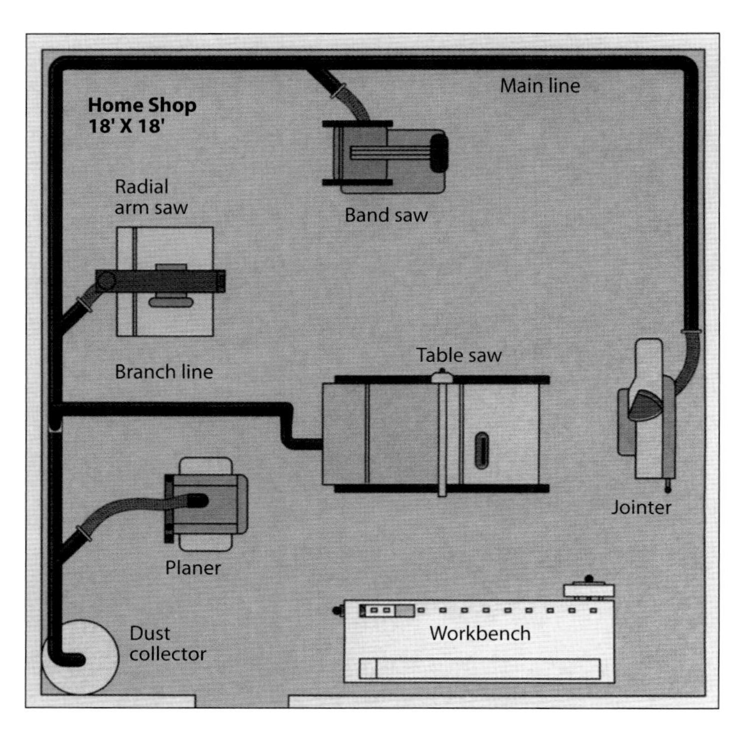

Designing a Shop for Efficient Dust Collection

Laying out a shop

The diagram at right illustrates a typical home shop layout. The power tools and dust collection system have been arranged for maximum dust collection efficiency. With the exception of the table saw, all the machines are situated on the perimeter of the work area. The ducting for the central dust collection system runs close to the walls. Despite requiring a relatively long main line, this design allows for short branch lines and minimal directional changes—both efficient arrangements. The space taken up by the dust collection system is minimized by placing the collector out of the way in a corner of the shop. The planer, probably the heaviest dust producer, is positioned closest to the collector to reduce strain on the system.

A central dust collection system requires a selection of fittings to route and join lengths of duct and dust hoods. The inventory below illustrates the elements of a typical dust collection system. If you run the main line along the ceiling, you can secure it in place with wire straps nailed to furring strips mounted between the joists.

Fittings directly affect the efficiency of the system, so choose them carefully. As a rule, gentle curves are better than sharp turns, so use Y fittings instead of Ts for branch connections, wherever possible. A blast gate should be located at each branch outlet to seal ducts when they are not being used, thereby increasing air flow to the machine in use. Hoods, whether commercially made or shop-built, should be positioned as close as possible to the source of the dust.

You have a choice of methods for connecting ductwork. Many ducts and fittings can be friction fit and secured with adjustable hose clamps. Duct tape can also effectively join plastic pipe. To ensure smooth air flow, metal ducts should be joined with rivets, rather than screws or bolts.

Once you have completed the layout of your system and selected the type of duct you will use, it is time to calculate your dust collection needs and select a collector. This involves determining the requirements of the heaviest dust collection task your system must handle. This usually will be the sum of system losses and the air volume demanded by the machine most distant from the collector. Purchase a collector with slightly more capacity. System losses are caused by such inefficiencies as bends in the line, corrugated ducting, leaks, and hoods without flanges. Use the charts and information on page 64 to size and select a collector.

Elements of a Dust Collection System

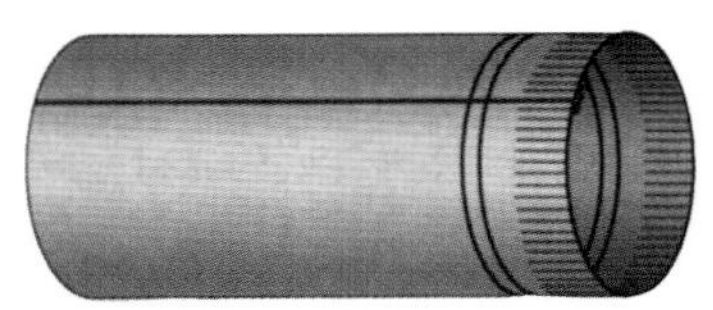

Metal duct
Standard dust collection pipe; available in wide range of diameters.

Reducer
Connects duct of different diameters; also used to increase suction in system or join a branch line to hood.

Adapters
Joins non-standard hose and duct to standard dust collection hose; also used to attach collection hose to factory-installed ports on stationary machines.

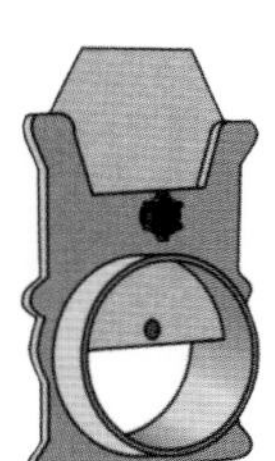

Blast gate
Pipe fitting with sliding gate which is opened or closed to direct dust collection air flow to a particular machine.

Splice
For joining two lengths of duct.

Hose clamp
Slotted metal band and screw used to join two lengths of duct or hose.

T connector
Connects two ducts at 90° angle.

PVC pipe
Plastic pipe for small shop dust collection systems; available in different diameters and wall thicknesses.

Hood
Dust-capturing device positioned close to source and connected to branch line.

Elbow
Attaches to duct to change direction of line.

Y connector
Joins two ducts at 30° or 45° angle.

Calculating Dust Collection Needs

Determining static pressure loss

Dust collectors are rated by their ability to move a certain number of cubic feet of air per minute (cfm) against a specific static pressure. The most important variable to keep in mind when choosing a dust collector for your shop is static pressure loss, which is a measure of the friction air encounters as it passes through a duct. The longer the ducting and the more numerous the system losses, the greater the static pressure loss. To determine the size of collector you need, calculate the static pressure loss for the heaviest collection task in the shop. In the diagram on page 62, it is the jointer. The following calculations are based on it. For your own shop, you may need to do the calculations for a few machines—those farthest from the collector and at the end of branch lines—and choose a collector based on the highest result you obtain.

Start with chart 1 *(right, top)* to calculate the equivalent length of the ducting running to the machine. In our example, there are 45 straight feet of smooth 4-inch-diameter duct and two 90° curved elbows. The equivalent length therefore is: 45 feet + 20 feet = 65 feet. Then use chart 2 *(middle)* to determine the exhaust requirements in cfm of the machine; for the jointer, it is 300 cfm.

Finally, use chart 3 *(bottom)* to determine static pressure loss for dust collection at the machine. Choose from either the third or fourth column of the chart depending on whether the machine is joined to a main-line duct (3500 feet per minute of air velocity, or fpm) or a branch line (4000 fpm). In this example, a 300-cfm machine connected to a 4-inch-diameter main line has a static pressure loss of .05 inches per foot. Thus the static pressure loss for this jointer is: 65 feet × .05 inches/foot = 3.25 inches. Add two inches for unmeasured losses like air leaks and the value rises to 5.25 inches. The shop on page 62 would need a collector with a 300 cfm rating at 5.25 inches of static pressure. A system 20 percent larger would allow for future expansion.

(1) Equivalent Length of System Elements

Duct or Fitting	Equivalent Length, in Feet
Smooth-wall pipe	Actual length
Corrugated pipe or hose	1.5 × actual length
Unflanged duct, hose, or hood connections	10
90° sharp elbow	20
90° curved elbow	10
90° hose bend	10
45° curved elbow	5
45° hose bend	5
Side leg of 90° T	20
Side leg of 45° Y	5

(2) Air Exhaust Volume Requirements for Machines

Machine	Cubic Feet per Minute (CFM)
Jointer (4–12")	300
Disc sander (up to 12")	300
Vertical belt sander (up to 6")	350
Band saw (up to 2" blade)	400
Table saw (up to 16")	300
Radial arm saw	350
Planer (up to 20")	400
Shaper (½" spindle)	300
Shaper (1" spindle)	500
Lathe	500
Floor sweep	350
Drill press	300
Jig saw	300

(3) Static Pressure Loss per Foot of Duct at 3500 and 4000 FPM

CFM	Duct diameter	3500 fpm	4000 fpm
300	4"	.05 in/ft	.07 in/ft
350	4"	.05 in/ft	.07 in/ft
400	4"	.05 in/ft	.06 in/ft
500	5"	.04 in/ft	.06 in/ft

Dust Collectors

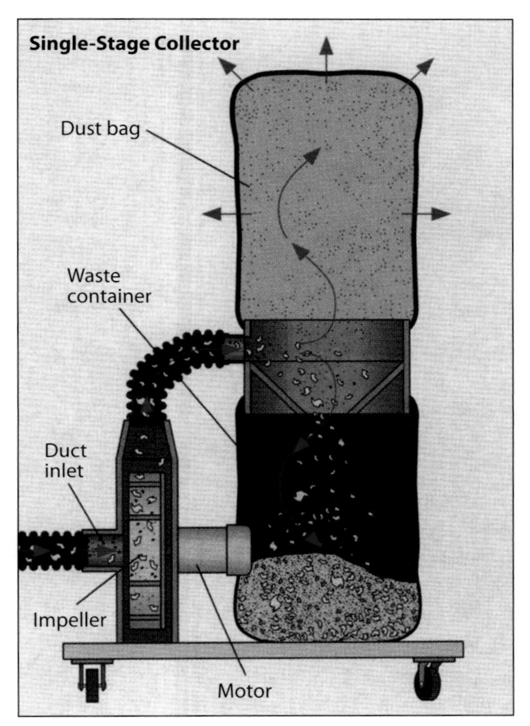

Choosing between single- and two-stage collectors

Two basic types of dust collectors are available for home workshops: single- and two-stage machines. In single-stage collectors *(above, left)*, debris- and dust-laden air is drawn through an impeller, where cyclone action deposits heavy dust and debris into the waste container below while the lighter dust rises to the dust bag. Single-stage collectors are relatively loud and the dust and debris tend to wear out the bag and impeller quickly. In two-stage collectors *(above, right)*, the impeller is located above the inlet duct so the heavier particles drop into the waste container before any air passes through the impeller and bag. This is quieter, and reduces wear on the impeller and dust bag. Two-stage collectors are somewhat more difficult to clean.

Shop Tip

Electrical sweeps for right-angle joints

If you use PVC pipe for your dust collection system, substitute 90° electrical sweeps *(far right)* for conventional 90° elbows *(near right)* to reduce friction and increase the efficiency of your system. Available at electrical supply houses, these fittings feature a 24-inch radius curve, which is much gentler than the 5- to 10-inch radius curve of standard elbows.

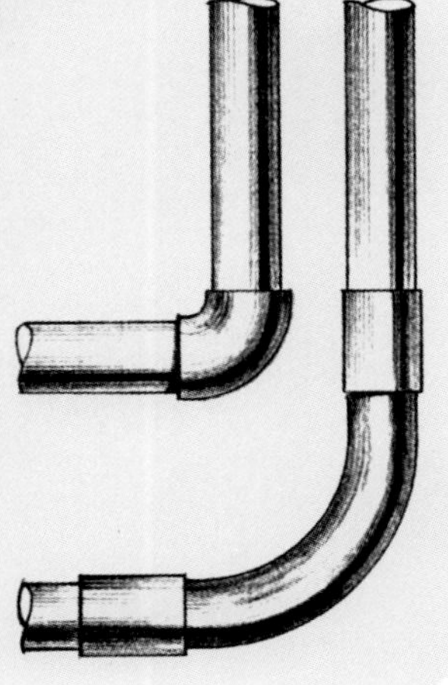

Dust Hoods

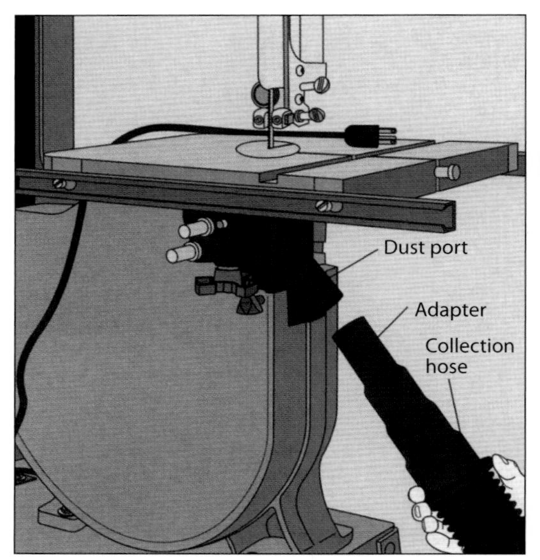

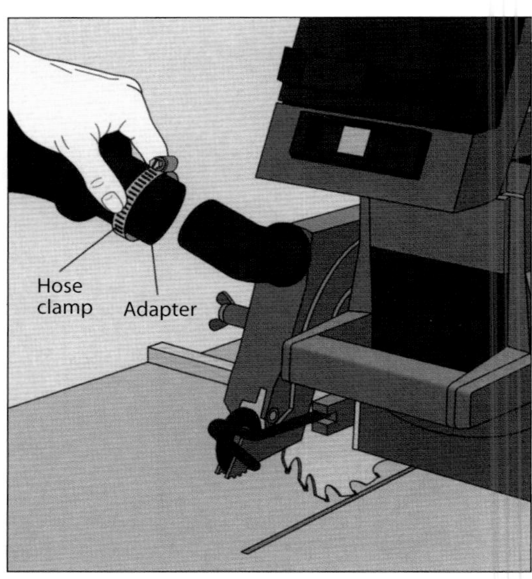

Connecting a dust collection system to tools with dust ports
Use a commercial adapter to attach a collection hose to a machine dust port. The adapter should be sized to friction-fit with the collection hose at one end and slip over the dust port at the other, as shown on the band saw *(above, left)*. For the radial arm saw, a hose clamp is used for reinforcement *(above, right)*.

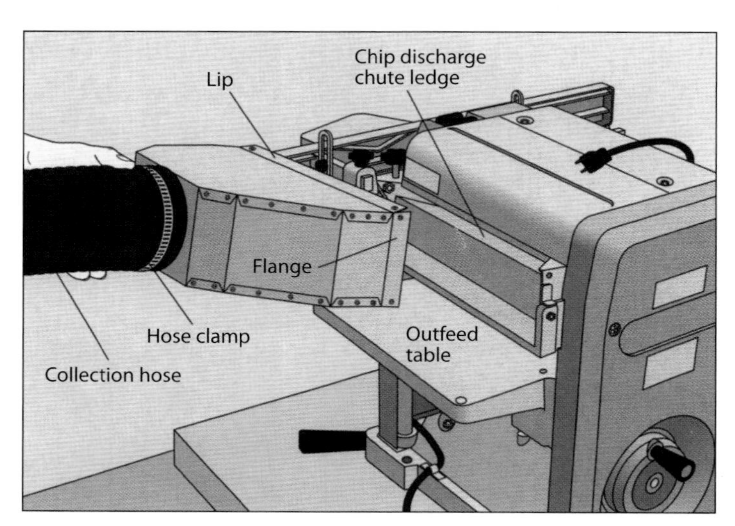

Hooking a planer up to the system
A hood like the one shown at right can be custom-built to capture most of the dust generated by your planer. Make the hood from galvanized sheet metal, cutting the pieces with tin snips. Leave tabs where the pieces overlap so they can be pop riveted together. Make flanges on the sides to improve the seal and a hole in the back for the dust collection hose; you will also need to create a lip along the top to connect to the ledge of the planer's chip discharge chute. Use an adapter to join the hood to the hose, inserting one end in the hole in the hood and the other end in the hose; reinforce the connection with a hose clamp. Fasten the lip of the hood to the planer with sheet metal screws.

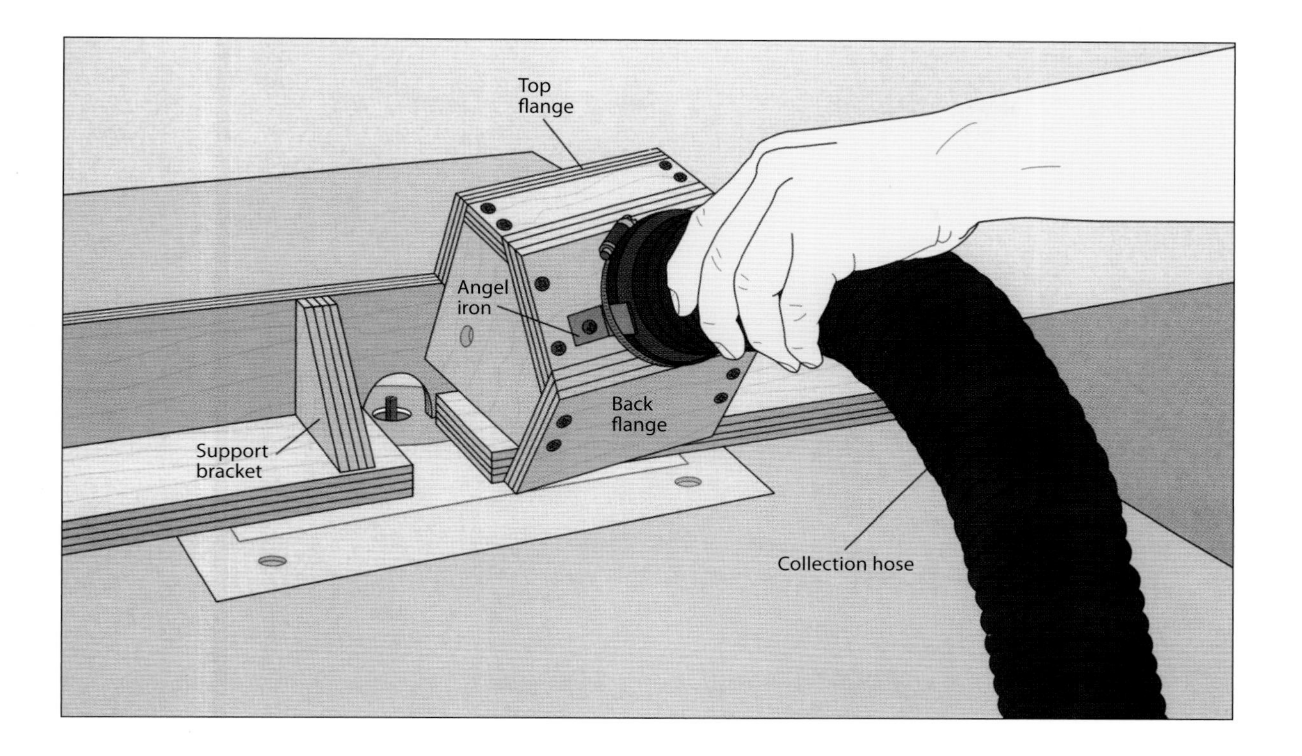

Shop Tip

Adapting standard sheet-metal ducts as dust hoods

Commercial sheet-metal ducts can be modified to serve as efficient hoods for your shop's dust collection system. Some examples are shown here. Use tin snips to cut the duct to a shape that suits the tool at hand. The duct should fit snugly around the chip discharge port or dust spout of the machine. Screw it in place with sheet metal screws.

Connecting a collection hood to a router table

A hood attached to the fence of a router table will collect most of the dust produced by the tool. Cut the hood from ½-inch plywood, sizing it so the sides hug the outside edges of the fence's support brackets. The bottom edge of the back flange should rest on the table; the top flange should sit on the top edge of the fence. Before assembling the pieces of the hood, cut a hole through the back for the collection hose. Also bore holes for screws through the sides and screw angle irons to the back so that their inside edges are flush with the opening for the hose. Screw the hood together, then fit the collection hose in the back. Use a hose clamp to secure the hose to the angle irons and position the hood on the fence *(above)*. Screw the sides of the hood to the fence brackets.

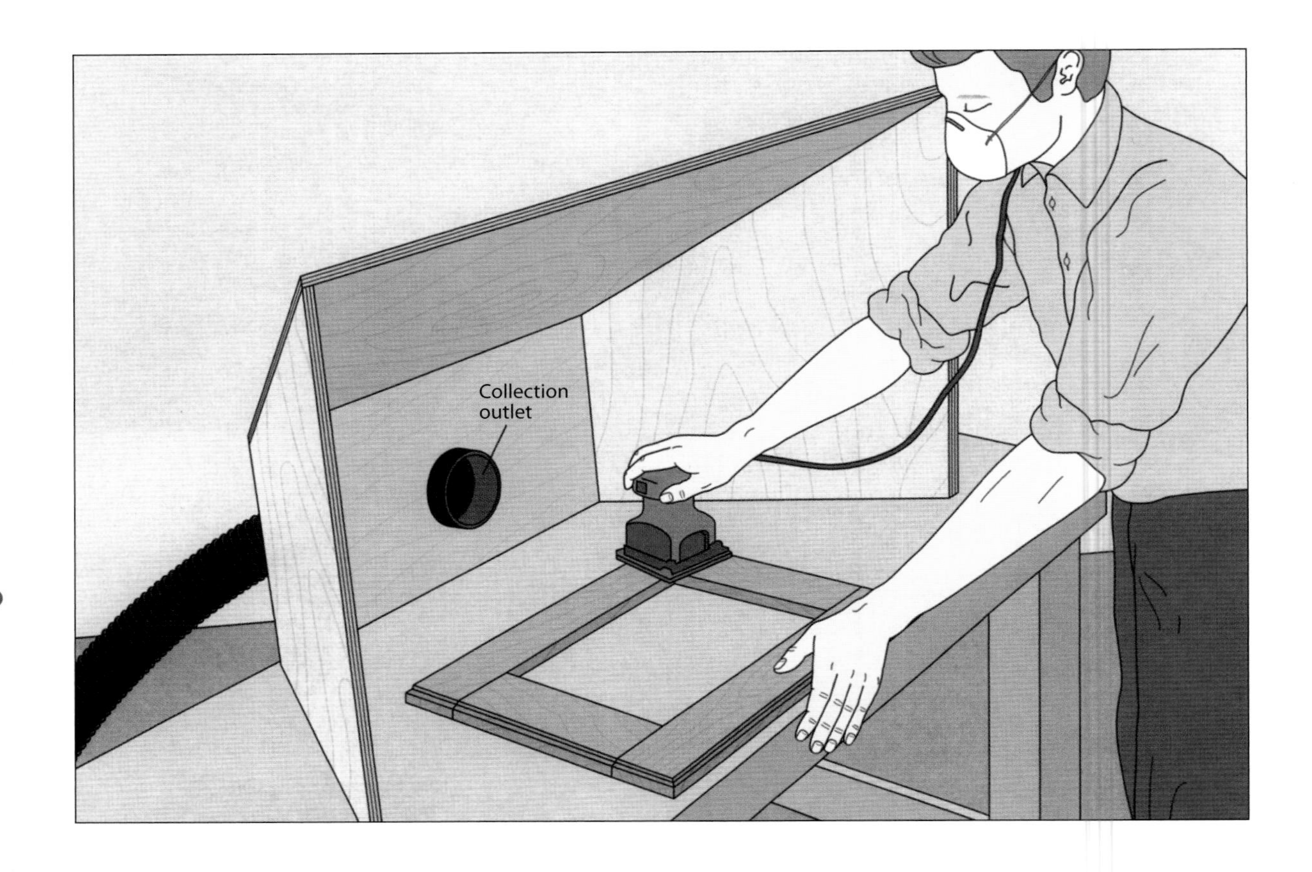

Setting up a shop-made sanding station
To reduce the amount of dust generated by power sanding, build a portable stall that fits on a table or workbench. Cut the back, top, and sides from ½- or ¾-inch plywood. Taper the top edges of the sides to create a comfortable, open working space, like the one shown above. Cut an outlet in the back of the station for a dust collection hose or branch duct. Assemble the station with screws. Position the sanding station securely on your work surface; attach the collector hose to the outlet. Turn on the collector before you begin a sanding operation.

Shop Tip

Shop-made blast gate
To fashion an inexpensive blast gate for a plastic duct, saw halfway through the pipe. Cut a gate from plywood or hardboard to fit in the kerf. Saw a semicircle in one half of the gate the same size as the inside diameter of the pipe; the other half should protrude from the kerf to form a handle. To seal the slot when the machine is in use, cut a sleeve from the same size of pipe with a diagonal slit to allow it to slide over the kerf.

Portable Dust Collection

A central dust collection system may sound like overkill to the craftsman with a small home shop. Although such systems are generally more efficient than independent collectors, they can be costly and consume considerable space. If your shop area is restricted, and only one machine will be operated at a time, consider a portable dust collector.

Many home woodworkers will find that a shop vacuum, although not ideal, can do a satisfactory job most of the time. Shop vacuums are designed to move a small volume of air at high velocity through a small-diameter hose. Dust collectors, on the other hand, move a large amount of air at a lower speed. A shop vacuum dust hood, therefore, should be positioned very close to the tool. Larger chips will tend to clog vacuum hoses, requiring frequent cleaning.

If there is no dust collection system—portable or central—in your shop, try the methods described starting on page 70 to control airborne dust. These methods are also effective supplements to collectors that suck up a majority of shop dust, but still leave some particles floating in the air.

Although you can hook up a band saw to a central dust collection system, another solution is to attach it to a portable shop vacuum. The vacuum's hose can often be slipped around an existing port on the machine using a simple commercial reducer.

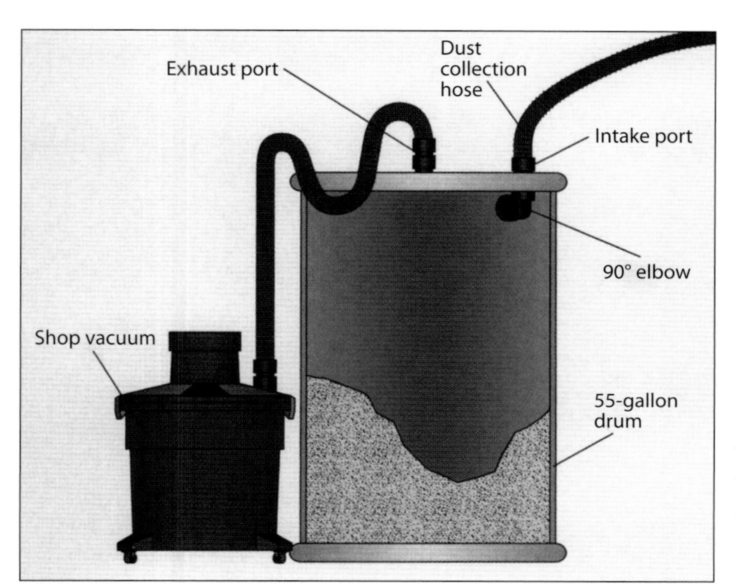

An Auxiliary Portable Dust Collection System

Expanding a dust collector's capacity

You can more than double the capacity of your portable dust collector or shop vacuum by attaching a 55-gallon drum or a large plastic barrel as a mid-stage collector. Install plastic intake and exhaust ports on the drum as shown at left and mount a hose to the intake port on the drum to collect wood dust and chips. The 90° elbow on the intake port will create a cyclone effect inside the barrel, forcing chips and heavier sawdust against the walls of the barrel. Lighter dust will be drawn through the exhaust port into the shop vacuum or dust collector. For easy assembly and disassembly, use pipe fittings that form a friction fit.

Controlling Airborne Dust

Setting up positive-pressure ventilation
To maintain clear air in a shop when you are generating a great deal of airborne dust or chemical fumes, set up a positive-pressure ventilation (PPV) system. Open all the windows in the shop and position a fan outside the door as shown at right so that the airflow it produces will envelop the doorway. The stream of air will follow the path of least resistance—through the door and shop, and out the windows, clearing airborne dust and fumes quickly. PPV has some limitations, however. The system will only function properly if the window openings are large enough to handle a sufficient volume of air. Also, the rest of your home must be well sealed off from the shop. A more permanent alternative to PPV can be fashioned by mounting an explosion-proof exhaust fan in a shop window. Set up to pull air out of the room, the fan will create negative pressure, expelling fumes and dust in larger volumes than is possible with PPV.

Shop Tip

Vacuum screening ramp
For cleaning dust off the shop floor, build a wedge-shaped screening ramp from ½-inch plywood. Before assembling the pieces, cut an inlet port in the back to fit a dust collection hose and five rows of 2-inch-diameter holes through the top. When dust and chips are swept up onto the ramp, smaller particles will fall through the holes and continue on to the collector. Larger refuse will remain on the ramp for easy disposal.

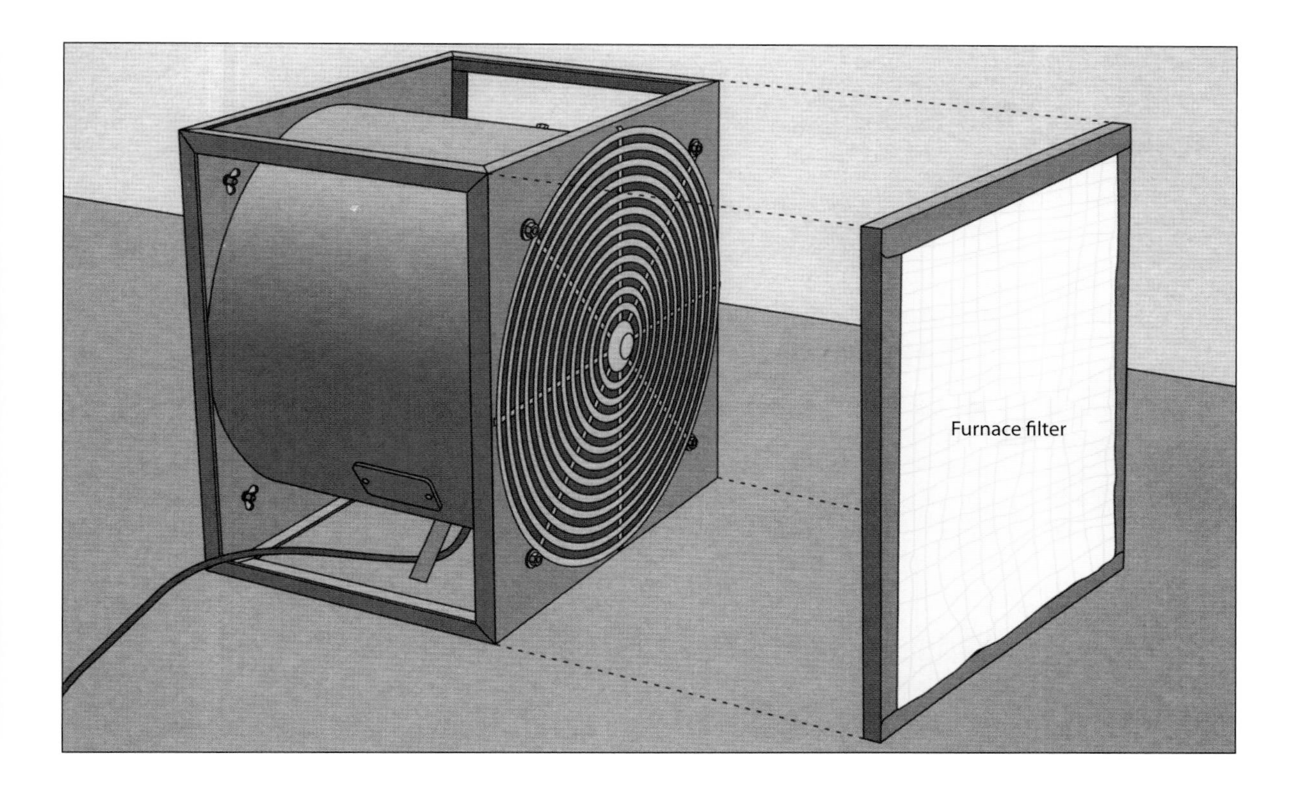

Shop Tip

Panty hose shop vacuum filter
Used panty hose can serve as an inexpensive alternative to replaceable shop vacuum dust filters. Fit the waist band around the foam filter sleeve on the underside of the motor housing of the vacuum and knot the legs. Slide the retaining ring around the panty hose to secure it in place.

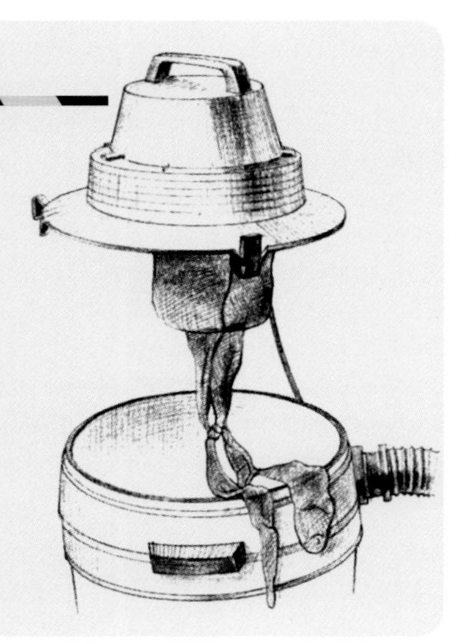

Filtering shop air
Another quick and easy method of ridding the shop of airborne dust uses a furnace filter on the back of a portable room fan *(above)*. When the fan is turned on, suction will hold the filter in place and draw dust out of the air. The dust will remain on the filter, which can then be brushed off outside or vacuumed.

Storage

Clutter is the woodworker's enduring enemy. Whether your workshop is shoehorned into the corner of the basement or spread out in a two-car garage, it no doubt accumulates things at an astonishing rate: Lumber, plywood, saws, saw blades, drills, drill bits, planes, clamps, chisels, files, grinders, screwdrivers, punches, wrenches, and hammers are just a few of the hand and power tools that must be conveniently available when needed—and out of the way when not.

Add to these the lumber scraps, locks, hinges, screws, nails, spare parts, and containers half full of finishes—all sure to be invaluable some day very soon—and you may have the makings of a monumental storage problem.

Adequate workshop storage should accomplish two goals: Tools and materials should be kept within easy reach of each operation, and the storage devices should encroach as little as possible on work space. No matter what your particular needs, you should find a number of storage ideas that conserve space in this chapter.

In evaluating your own storage options begin by taking two inventories: one of your tools and materials and the other of unused space—corners, spaces under the stairs, between wall studs, below the workbench, and between the ceiling joists. Lumber and plywood are best stored on racks that prevent warping and water damage, keeping the wood out of the way and easy to reach. Shelves, drawers, and cabinets are the most convenient and economical spaces for storing tools. There are a variety of commercial storage devices on the market, but you can build a tool cabinet customized to your needs easily and inexpensively *(page 81)*. The design shown on page 84 fulfills two needs in one: a storage cabinet that folds down and serves as a sturdy work surface. Hardware can be sorted in drawers, subdivided into separate compartments or, for greater visibility, in glass containers. For tools like clamps that are used all over the shop, consider a wheeled rack *(page 88)*.

No matter what devices and techniques you choose, you may find that proper storage not only provides more space and convenience, but conveys a sense of order and purpose that will make your shop an even more pleasant and productive place to work.

Flammable products like lacquers, shellacs, and paint thinners require special attention. Storing these items in a double-lined, explosion-proof steel cabinet is one sensible solution.

Whatever its size, a tool chest can serve as a cabinetmaker's calling card. This portable carver's chest keeps tools organized, safe from damage, and within easy reach.

Storing Wood

Properly stored lumber and plywood are not only kept out of the way but straight and dry, too. For most shops, this involves storing lumber in racks that hold the wood off the floor. Wood shrinks and expands according to the amount of humidity to which it is exposed. A wet floor can warp lumber and delaminate some plywoods. The lumber racks featured in this section are easy and inexpensive to build; you should be able to find a suitable design and adapt it to your needs.

If you have the space, you can set up an end-loading lumber rack like the one shown on page 75. Such a system is relatively easy to construct but you will need a wall twice the length of your lumber to allow for loading and unloading. If space in your shop is at a premium, consider a front-loading rack like the one shown on page 78. If versatility is needed, examine the rack on page 76, which allows you to store boards both horizontally and vertically. Avoid using Z-shaped brackets; they waste too much space.

The typical shop can stock hundreds of pounds of lumber, so it is crucial to anchor your rack firmly—to at least every second wall stud or floor joist.

Make the most of spaces that you would not ordinarily consider as prime storage areas. If your ceiling is unfinished, nail furring strips across the joists for handy shelving to store short stock and dowels.

Every item in a workshop demands its own storage method. The dowel rack at left, built from ¾-inch plywood, 1-by-4 stock and 6-inch-diameter cardboard tubes, sorts different sizes of dowels while taking up a minimum of floor space.

A Lumber Rack

Storing planks and boards

The storage rack at right features vertical supports screwed to wall studs. Cut from 2-by-4 stock, the supports buttress shop-made wood brackets, which hold up the lumber. You will need one support at each end of the rack, with an additional one every 32 inches along the wall. After bolting the supports to the studs, prepare the brackets by cutting the sides from ¾-inch plywood and the middle shelf piece from 2-by-4 stock 1½ inches shorter than the brackets. Angle the top edge of the sides by about 5° so the brackets will tilt up slightly *(inset)* and prevent the lumber from falling off the rack. Screw the middle shelf piece to the sides, then screw the bracket to the vertical supports.

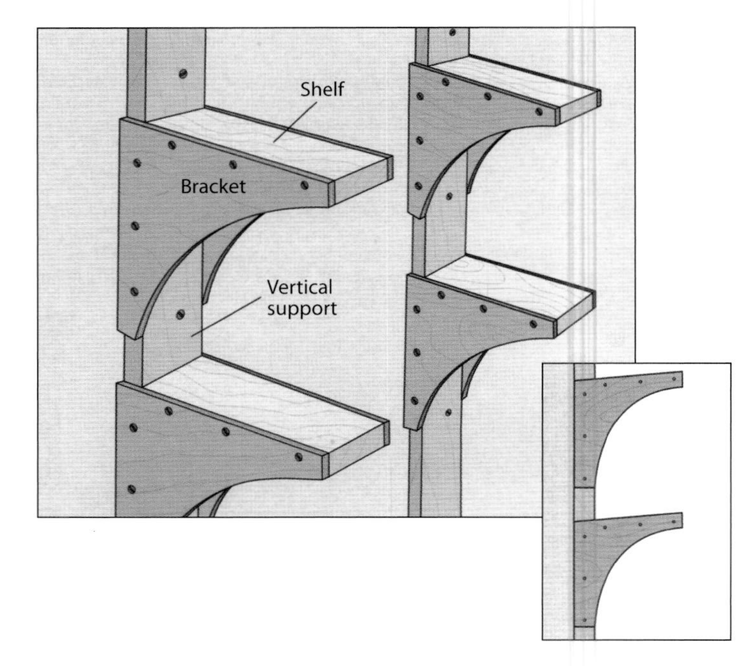

A Lumber-and-Plywood Rack

Fastening the rack to an unfinished wall

The rack shown below, made entirely of 2-by-4 stock, is attached to wall studs and ceiling joists. Lumber can be piled on the arms, while plywood is stacked on edge against the support brackets. You will need at least 8½ feet of free space at one end of the rack to be able to slide in plywood panels. Begin by cutting the triangular-shaped brackets and screwing them to the studs *(right)*. Cut the footings, slip them under the brackets and nail them to the shop floor. Next, saw the uprights to length and toe-nail their ends to the footings and the joists. Cut as many arms as you need, aligning the first row with the tapered end of the support brackets. Use carriage bolts to fasten the arms to the studs and uprights, making sure the arms in the same row are level. The rack in the illustration features arms spaced at 18-inch intervals.

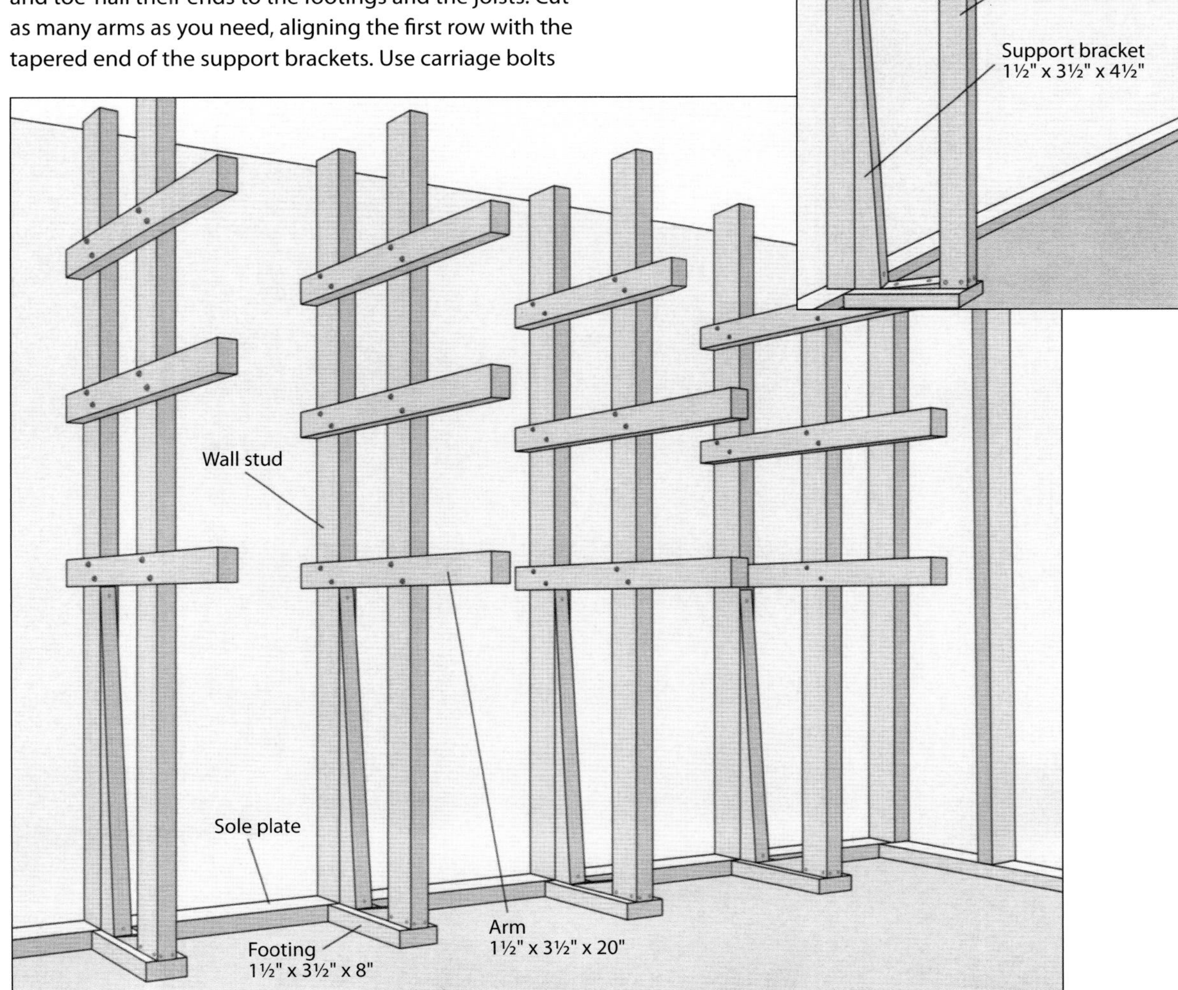

Adjustable Lumber Racks

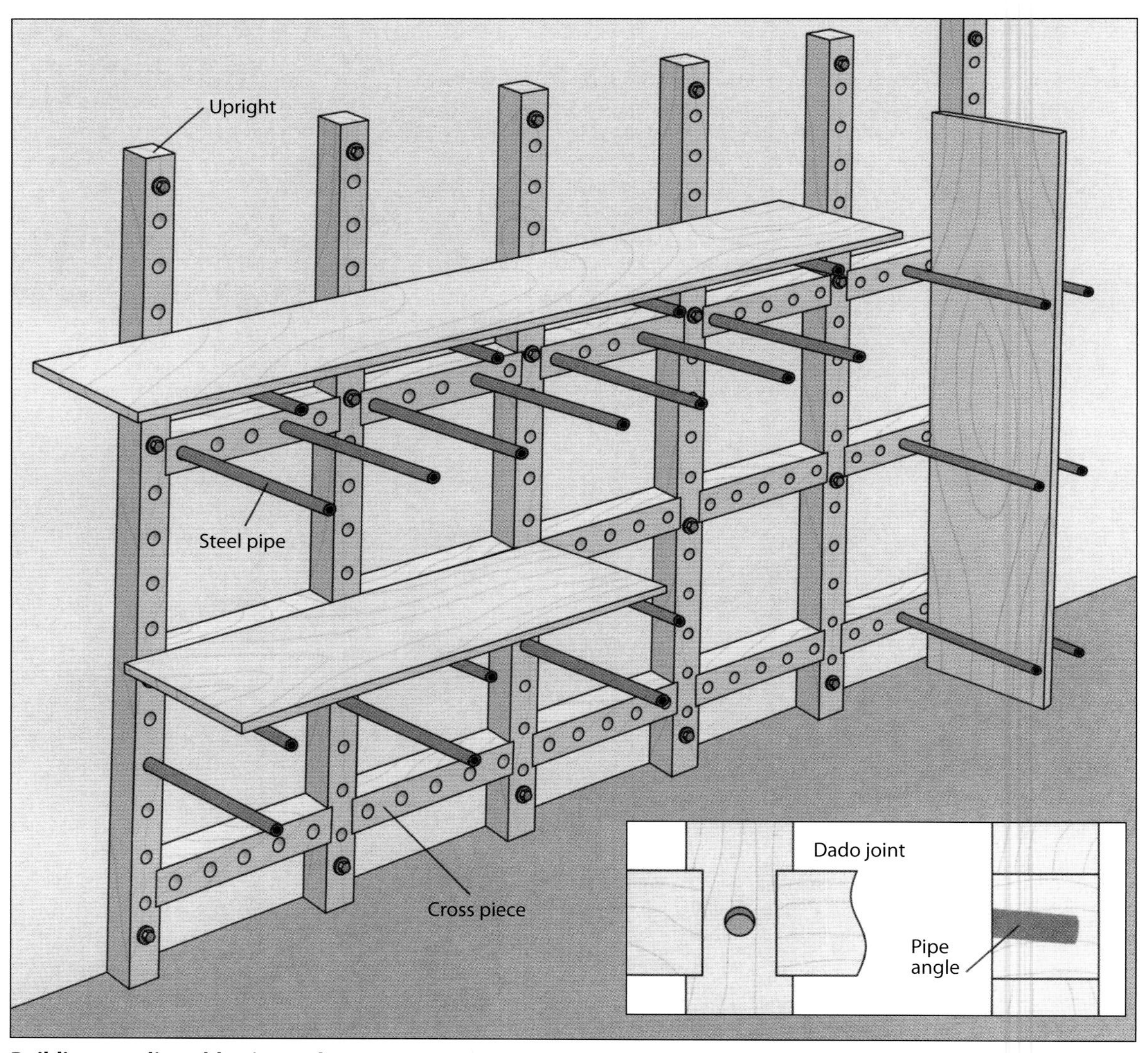

Building an adjustable pipe rack

The rack shown above, made of 4-by-4 stock and steel pipe, is attached to wall studs. The steel pipes should be roughly 24 inches long and ¾ inch in diameter. They can be inserted into any of the holes drilled into the vertical supports or crosspieces, allowing lumber to be piled on the pipes or stacked on end between them. Begin by cutting the uprights to length and mark each point on them where you want to locate a crosspiece. Cut dadoes in the sides of the uprights to accommodate the crosspieces *(inset)*, making sure all the crosspieces in the same horizontal row will be at the same level. Bore holes into the uprights and crosspieces for the pipes; drill the holes 3 inches deep and 6 inches apart, angling them by about 5° so the pipes will tilt up slightly. Bolt the uprights to the studs, then cut the crosspieces to length and tap them in between the uprights with a mallet. Fix them in place with glue or by driving in screws at an angle.

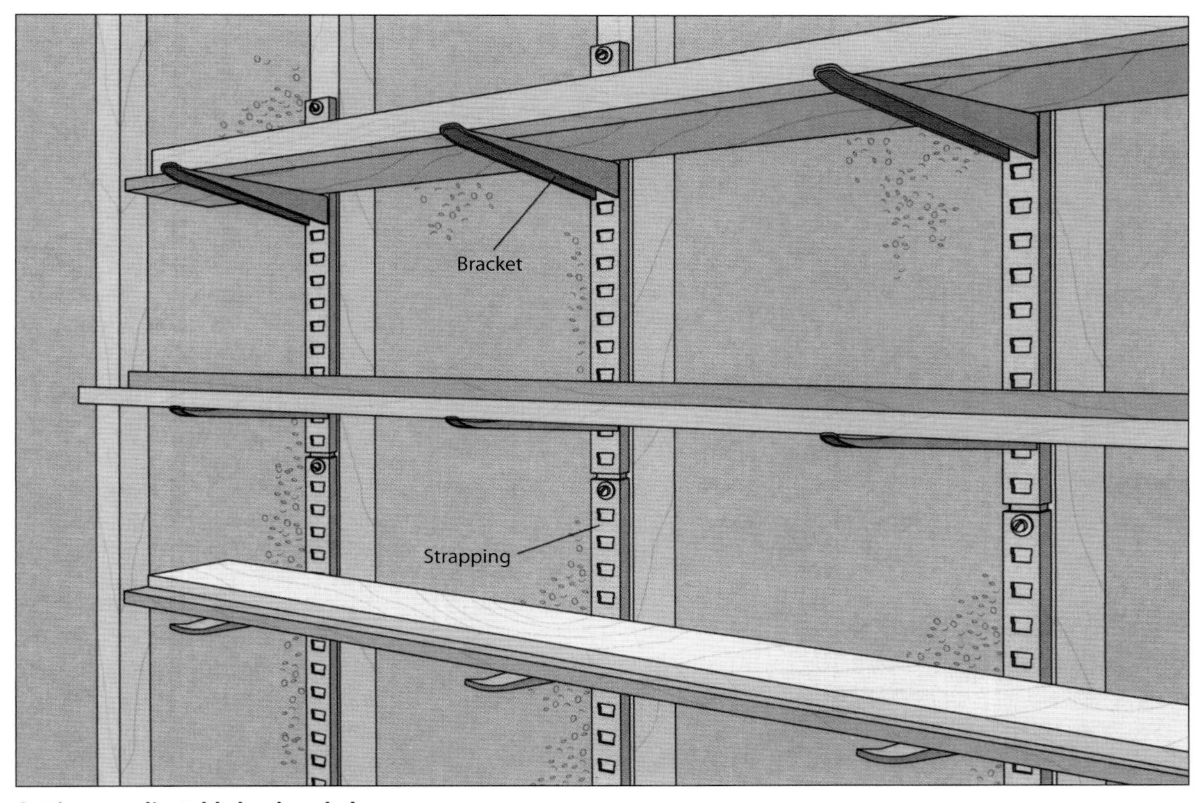

Setting up adjustable lumber shelves

A commercial lumber storage system like the one shown above consists of metal strapping and brackets that fit into holes in the strapping. The rack is similar to the wooden one on page 75, but because it is metal, this rack can typically support heavier loads. Bolt the strapping directly to the wall studs, or to vertical supports fastened to non-exposed studs. Make sure the straps are aligned laterally to allow you to position each row of brackets at the same height. For most applications, attach the brackets to the strapping about 24 to 36 inches apart vertically.

Lumber Storage Racks

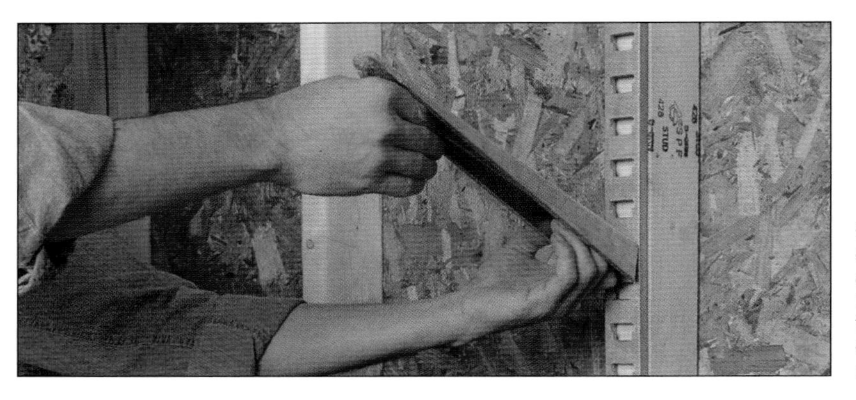

Commercial lumber racks, such as these cantilevered lumber shelves, are both adaptable and strong, making them ideal for the home worshop. Screwed to a concrete wall or to wall studs, they can be adjusted to various heights to suit your particular storage needs.

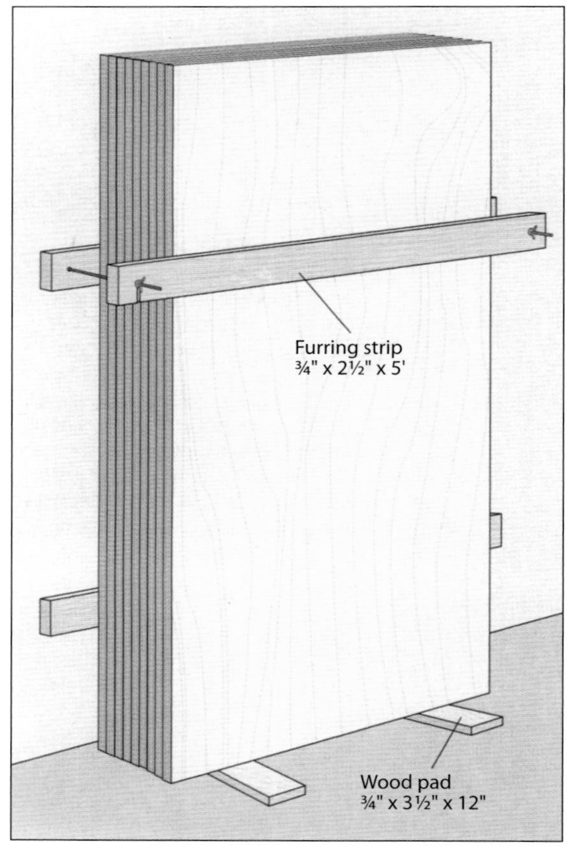

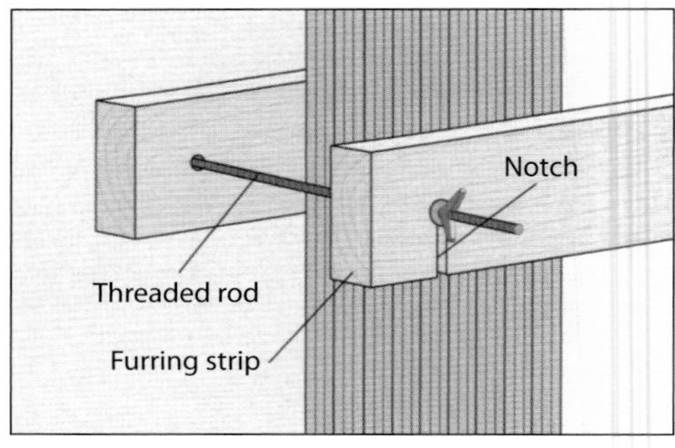

A Vertical Plywood Rack

For long-term storage, stacking plywood on end saves valuable shop floor space. The rack shown at left is built from furring strips, threaded rods, and wing nuts. Start by screwing two 1-by-3 furring strips to the studs of one wall, 2 and 5 feet from the floor; first bolt two threaded rods 4½ feet apart into the top strip. Cut a third furring strip and bore a hole through it at one end and saw a notch at the other end to line up with the rods. Both openings should be slightly larger than the diameter of the rods. Place two wood pads on the floor between the rods and stack the plywood sheets upright on them. Holding the third furring strip across the face of the last panel, slip one rod through the hole and the other into the slot. Put washers and wing nuts on the rods and tighten them, pulling the furring strip tightly against the plywood *(above)*. To remove a sheet from the stack, loosen the wing nuts and swing the furring strip up and out of the way.

Storing Tools and Supplies

"A place for everything and everything in its place." That time-worn adage is especially appropriate for the home workshop. From shelves and racks to tool chests and partitioned drawers, many devices will eliminate clutter while keeping tools and supplies easily accessible. A few methods are shown in the following pages.

For certain tools, particularly items that are valuable or dangerous enough to be out of the reach of children, wall-mounted boxes like those shown below are ideal. For a more traditional system of enclosed storage, you can build a tool cabinet or cupboard in the shop *(page 81)*. But not every storage device needs to be elaborate. As shown on page 86, suspending a tool from a fastener driven into a wall can work just fine.

Shelves are an ideal tool storage option. This shop-built unit features grooves and cleats custom-cut to hang a panoply of tools in full view over a workbench.

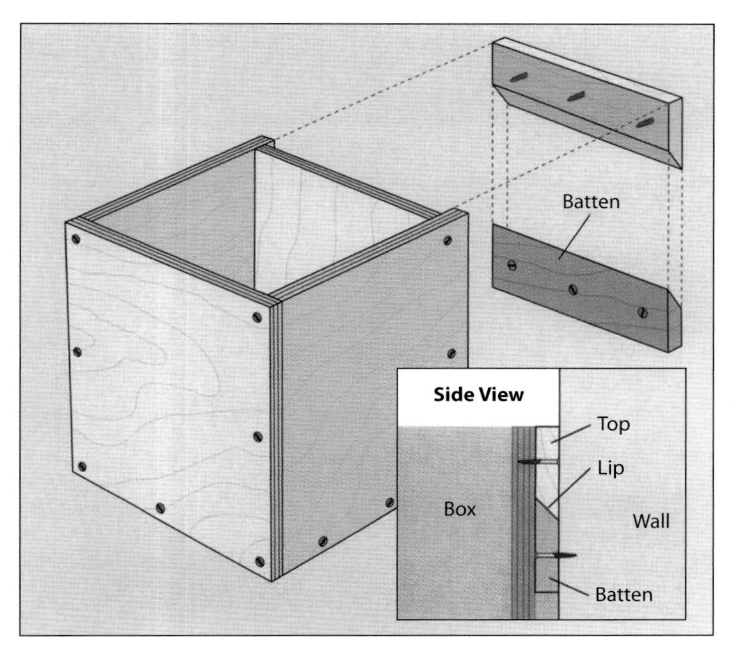

Wall Storage

The box at right can be hung securely on a shop wall and easily moved if necessary. Build it from ¾-inch plywood with a hinged top. To hang the cabinet on the wall, cut a 45° angle bevel down the middle of a 1-by-6, then crosscut the two pieces slightly less than the width of the box. Screw one of the pieces to the wall as a batten, with the bevel pointing up and facing the wall; anchor as many of the fasteners as possible in wall studs. Screw the other piece to the back of the box with its flat edge butting against the lip and the bevel pointing down and facing the back. The two pieces interlock when the box is hung on the wall *(inset)*.

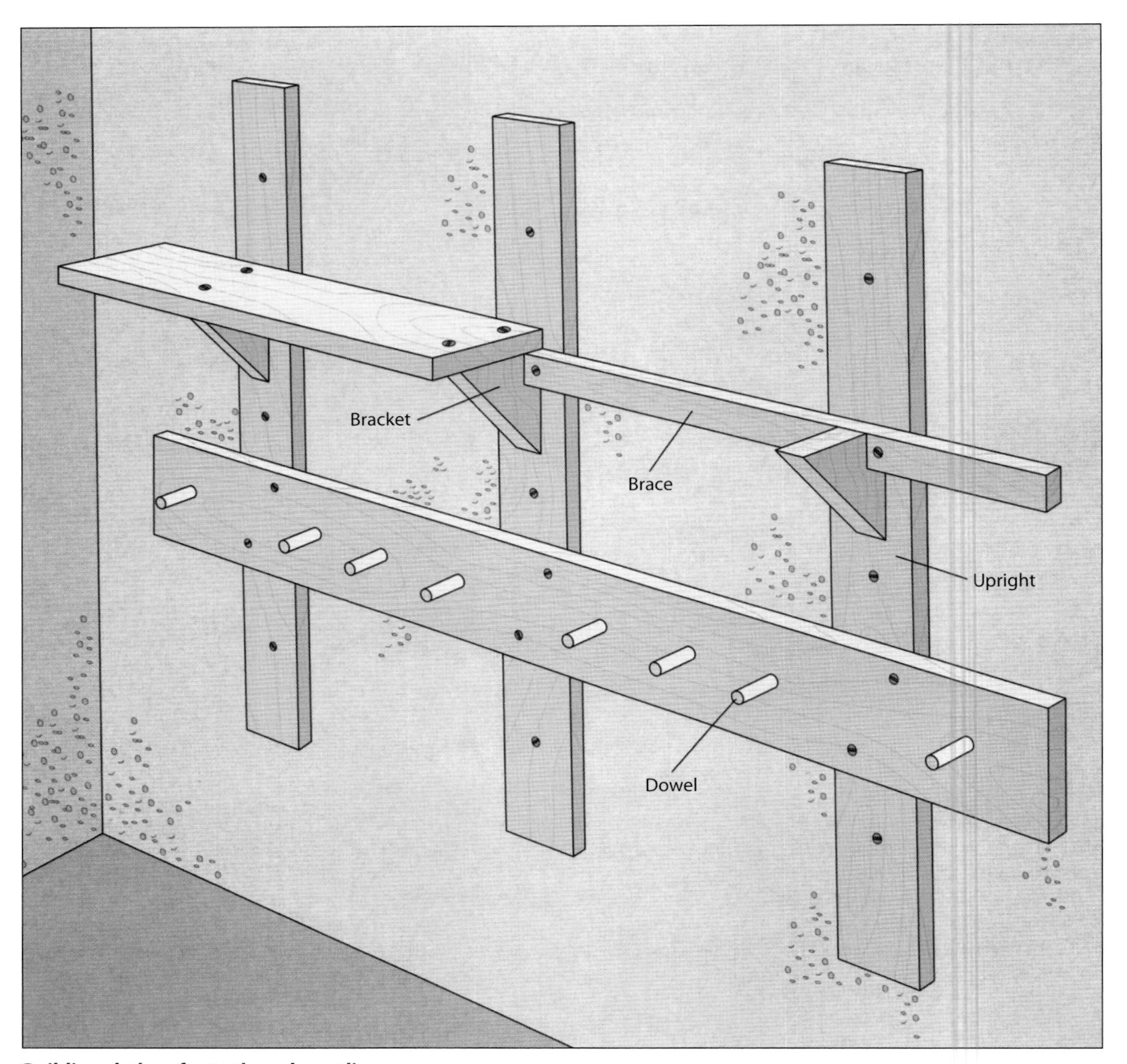

Building shelves for tools and supplies

The shelving shown above can be fastened to a concrete foundation wall using lead anchors or concrete screws. If your shop has a stud wall, anchor the uprights to the studs. Begin by cutting the uprights from 1-by-4 stock, allowing 32 inches of space between the uprights in order to fasten them to studs. To add a shelf, cut a brace from 1-by-2 stock and make a triangular bracket for every upright. Saw the brackets so that the side facing up is about 2 inches shorter than the width of the shelf. Also cut notches in the brackets for the brace, then screw them to the uprights, making sure they are all at the same height. To hang tools from the unit, screw a 1-by-6 across the uprights, then bore holes into the board for 1-inch-diameter dowel pegs. Glue the dowels in the holes.

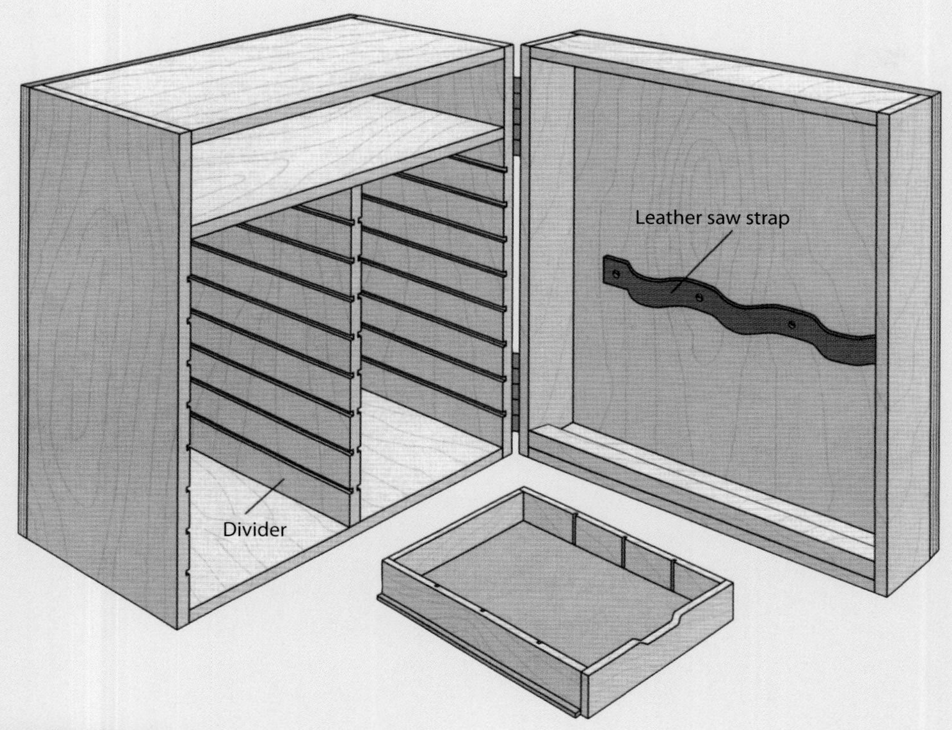

A Tool Cabinet

The tool cabinet shown above is handy for storing and organizing hand tools. Although the entire unit is portable, the drawers are removable, making it possible to carry around only the tools that are needed. Build the cabinet from either ¾-inch plywood or solid lumber. The size of the box will depend on your needs but 40 inches high by 30 inches wide by 15 inches deep is a good starting point. Position the divider in the center of the cabinet so that the spaces on both sides of it are equal, making the drawers interchangeable.

Cut the pieces to size, then prepare the sides of the cabinet and the divider for the drawers: Rout a series of ¼-by-¼-inch dadoes on one face of the sides and on both faces of the divider. Make the space between the dadoes equal to the height of the drawers, plus ¼-inch for clearance. Glue up the cabinet, shelf, divider, and door, using the joint of your choice. The cabinet in the illustration was assembled with plate, or biscuit, joints. Nail a leather strap to the inside of the door for hanging tools, add a wood strip to prevent small items from falling out, then attach the door to the cabinet with butt hinges.

Next, build the drawers. Saw the pieces to size, using ¼-inch plywood for the bottom; orient the panels so the grain of the face veneer runs from the front of the drawer to the back. Cut the sides slightly shorter than the depth of the cabinet if you are working with lumber, to allow for wood movement. Make the drawer front ¼-inch wider and cut a rabbet along its bottom edge to conceal the bottom, and notch the top edge for a handle. Cut dadoes in the sides for dividers. Glue up the drawers; the bottoms should extend beyond both sides by ¼ inch to form sliders that fit in the cabinet dadoes.

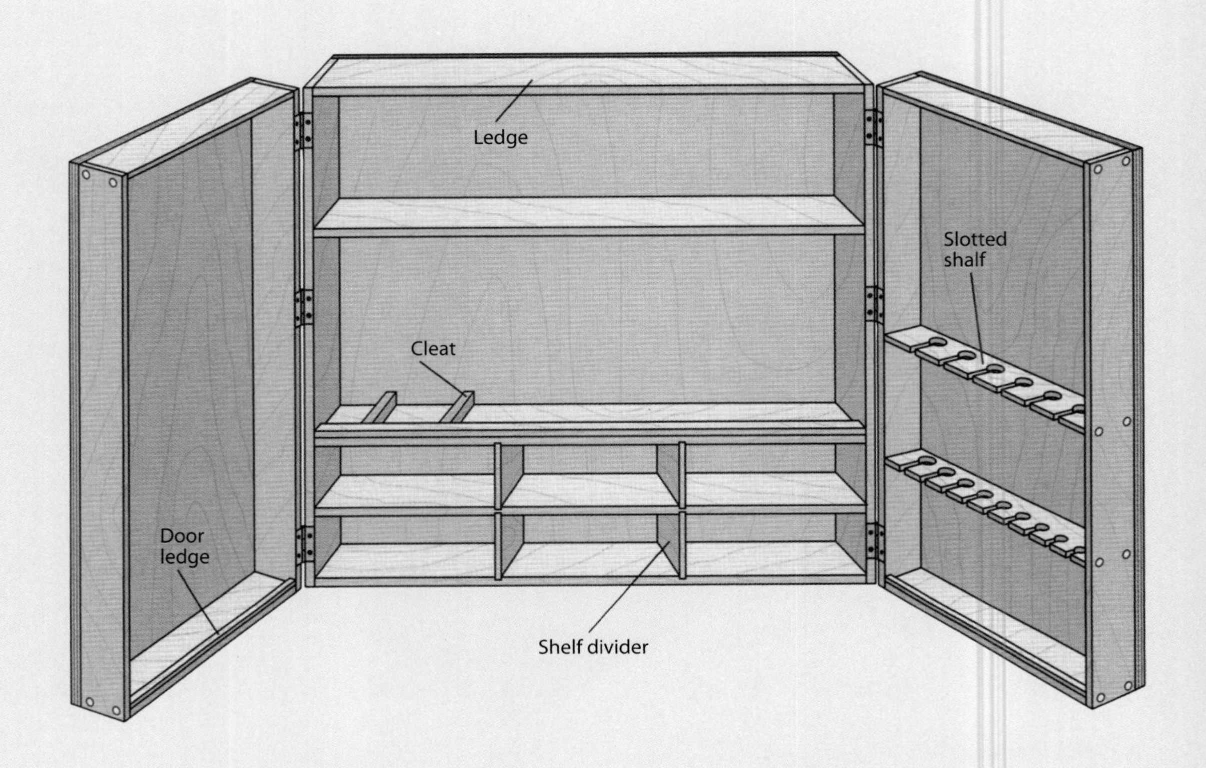

A Tool Cupboard

The cupboard above features twin doors for storing small, light tools like chisels and screwdrivers, as well as a large main compartment for bigger tools. Cut the components from ¾-inch plywood or lumber to the appropriate size, depending on the number of tools you own; the cupboard shown above is 48 inches square and 5 inches deep with 3-inch-deep doors. Next, assemble the cupboard using the joinery method of your choice. A through dovetail joint is one of the strongest and most visually pleasing options. But you could choose a method as simple as counterbored screws concealed under wood plugs, as shown above.

To help you install the shelves, lay the cupboard on its back and place the tools to be stored in their designated spots. Position the shelves accordingly and screw them in place. To keep supplies from rolling off a shelf or the bottom of the doors, glue a ledge along the front edge. If you want to subdivide a shelf, screw 1-by-1 cleats across it or install vertical dividers between the shelves.

Equip one or both doors with slotted shelves to hold tools like chisels and screwdrivers. Bore a series of holes slightly smaller than the tool handles, then saw a kerf from the edge of the shelf to the hole to enable you to slip in the blade. Screw the shelves to the door.

Hang the doors on the cupboard with butt or piano hinges. Use three butt hinges per door. Mount the cupboard to the wall above your workbench, if desired, by screwing it to the wall studs.

Keeping Tools Organized

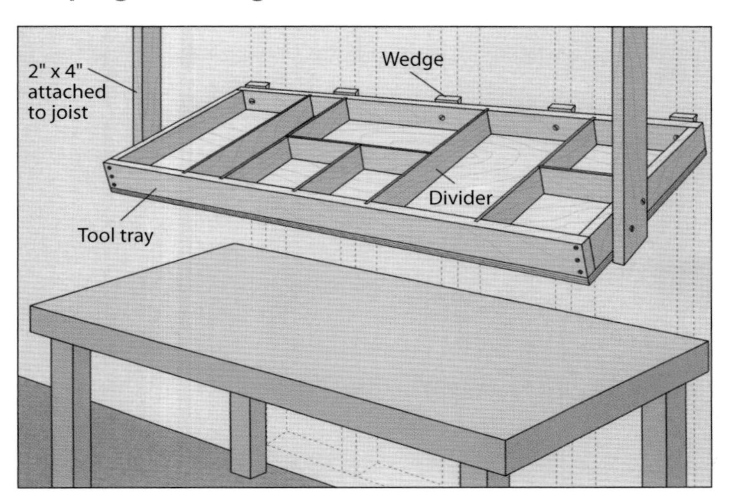

Storing hand tools

The tool tray shown above keeps different tools apart and similar ones together, helping to protect them while making a needed item easy to locate. The tray has the additional advantage of being suspended from overhead joists so that it takes up no valuable work space. Start by bolting two 2-by-4s to joists, spacing them to accommodate the tray. Cut off the bottom ends of the 2-by-4s at a convenient height. Next, build the tray, cutting the sides from ½-inch stock, and the bottom and the dividers from ¼-inch plywood. Cut dadoes for the dividers according to how you wish to group your tools, then screw the sides together and to the bottom. Glue the dividers in the dadoes and screw the sides to the 2-by-4s. Screw the back to wall studs, or, using lead anchors, to a concrete wall. If you plan to install the tray at an angle, as shown, drive the screws through wood wedges placed between the tray and the wall.

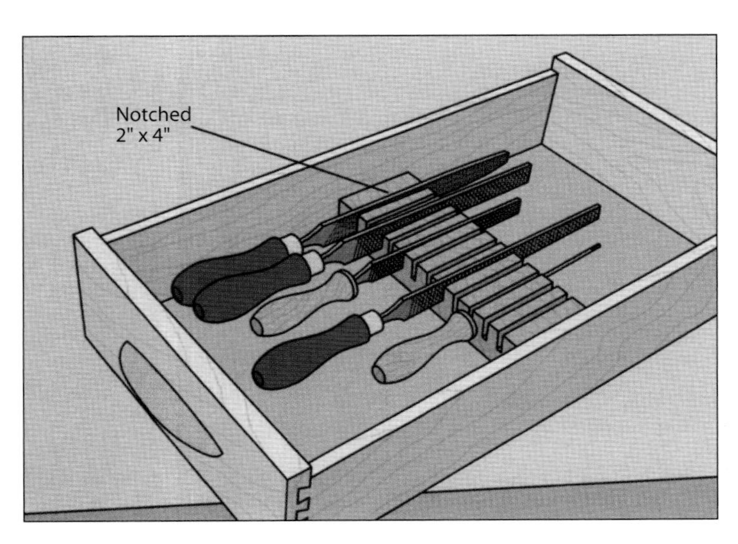

Adding tool-tray dividers

To protect tool edges in storage drawers, saw a 2-by-4 to a length equal to the space between the drawer sides. Then cut dadoes across one side of the board to hold the tools—in this case, narrow dadoes to accommodate file blades *(left)*.

A Fold-Down Workbench and Tool Cabinet

Ideal for small workshops, the storage cabinet shown below and opposite features a door that serves double-duty as a sturdy work surface that folds up out of the way when it is not needed. Mounted on a frame that is anchored to wall studs, the unit is built with an adjustable shelf and a perforated hardboard back for organizing and hanging tools as well as a work table supported by folding legs. The cabinet-bench can be made entirely of ¾-inch plywood, except for the legs and leg rail, which are cut from 2-by-4 stock; the 1-by-3 frame; the 1-by-4 hinge brace assembly; and the hardboard back.

Build the unit in three steps, starting with the frame, then making the cabinet section, and finally cutting and attaching the work table and legs. Referring to the cutting list for suggested dimensions, cut rabbets in the frame rails and stiles, then glue and screw them together. Next, screw the frame to the studs in your shop. Be sure to position the frame so that the work surface will be at a comfortable height, typically about 36 inches off the

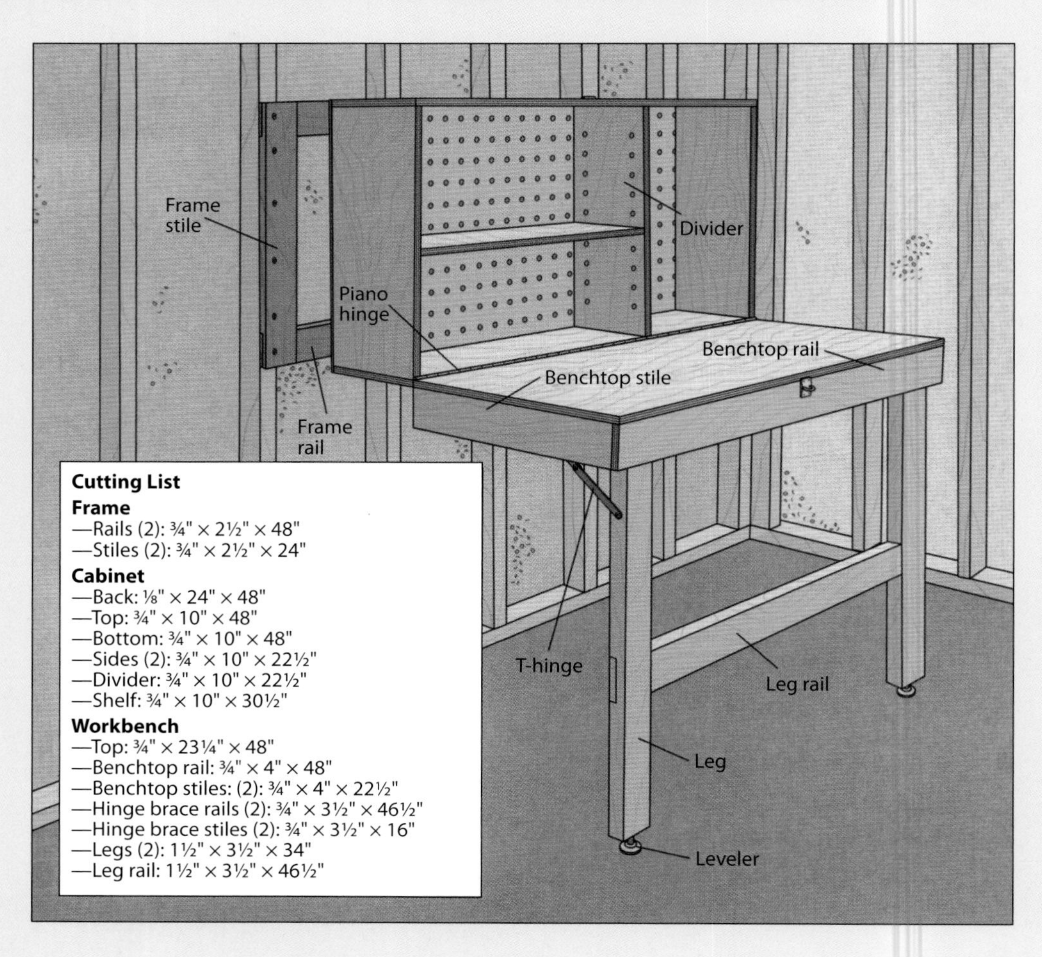

Cutting List

Frame
—Rails (2): ¾" × 2½" × 48"
—Stiles (2): ¾" × 2½" × 24"

Cabinet
—Back: ⅛" × 24" × 48"
—Top: ¾" × 10" × 48"
—Bottom: ¾" × 10" × 48"
—Sides (2): ¾" × 10" × 22½"
—Divider: ¾" × 10" × 22½"
—Shelf: ¾" × 10" × 30½"

Workbench
—Top: ¾" × 23¼" × 48"
—Benchtop rail: ¾" × 4" × 48"
—Benchtop stiles: (2): ¾" × 4" × 22½"
—Hinge brace rails (2): ¾" × 3½" × 46½"
—Hinge brace stiles (2): ¾" × 3½" × 16"
—Legs (2): 1½" × 3½" × 34"
—Leg rail: 1½" × 3½" × 46½"

floor. Now build the cabinet section, cutting the parts to size. Before assembling the pieces, bore two parallel rows of holes on the inside face of one side panel and the opposing face of the divider. Drill the holes at 1-inch intervals about 2 inches in from the edges of the panels. By inserting commercially available shelf supports in the holes, the height of the shelf can be adjusted to suit your particular needs. With the exception of the shelf, screw the parts together, then cut the hardboard to size and nail it to the cabinet. Fit the unit against the frame and use screws to attach the cabinet to the frame.

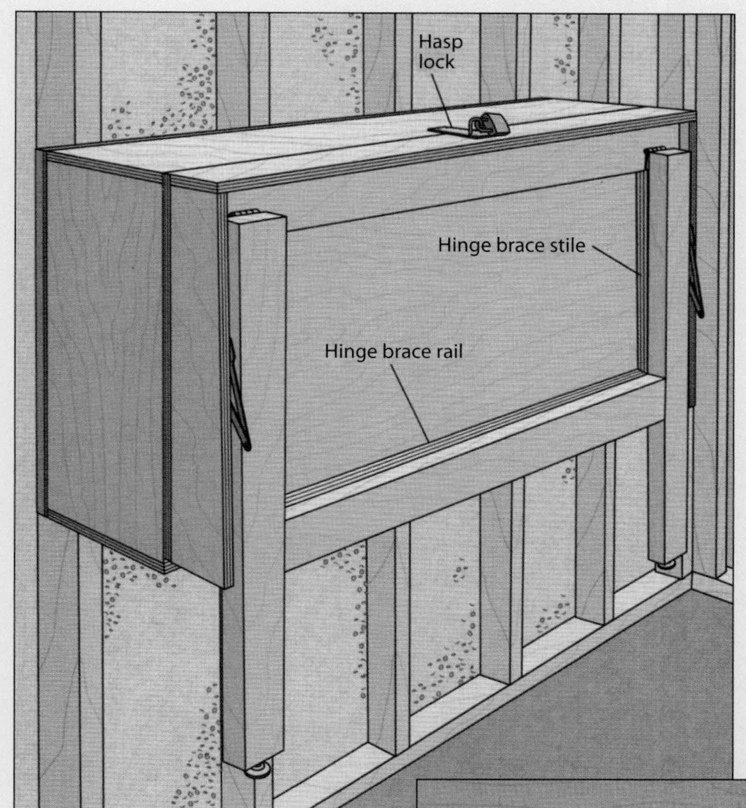

Saw the parts of the workbench to size, then screw the hinge brace rails and stiles and the benchtop rail and stiles to the underside of the benchtop *(above, right)*. Attach the benchtop to the bottom of the cabinet section with a piano hinge, making sure the two edges are perfectly aligned.

With the workbench folded down and held parallel to the floor, measure the distance from the hinge brace rail to the floor and cut the legs to fit. Attach the legs to the rail with hinges, then screw levelers to the bottom of the legs and adjust them as necessary to level the benchtop. Add a folding metal brace to each leg for added support, screwing the flat end of the brace to the hinge brace stile and the other end to the outside edge of the leg. Also cut a leg rail to fit between the legs and screw it in place. Finally, install a hasp lock, screwing one part to the top of the cabinet and the other part to the benchtop rail.

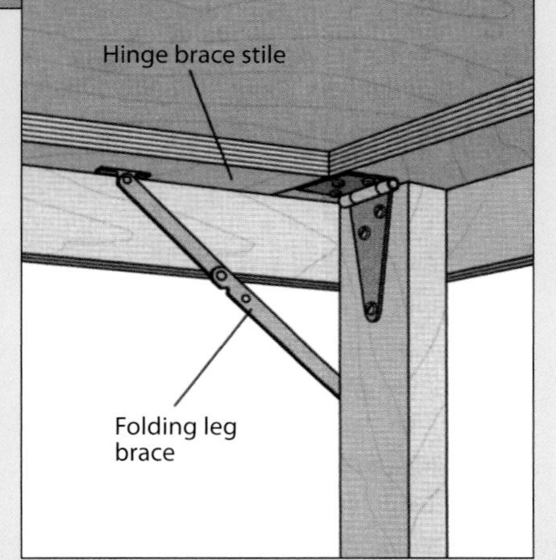

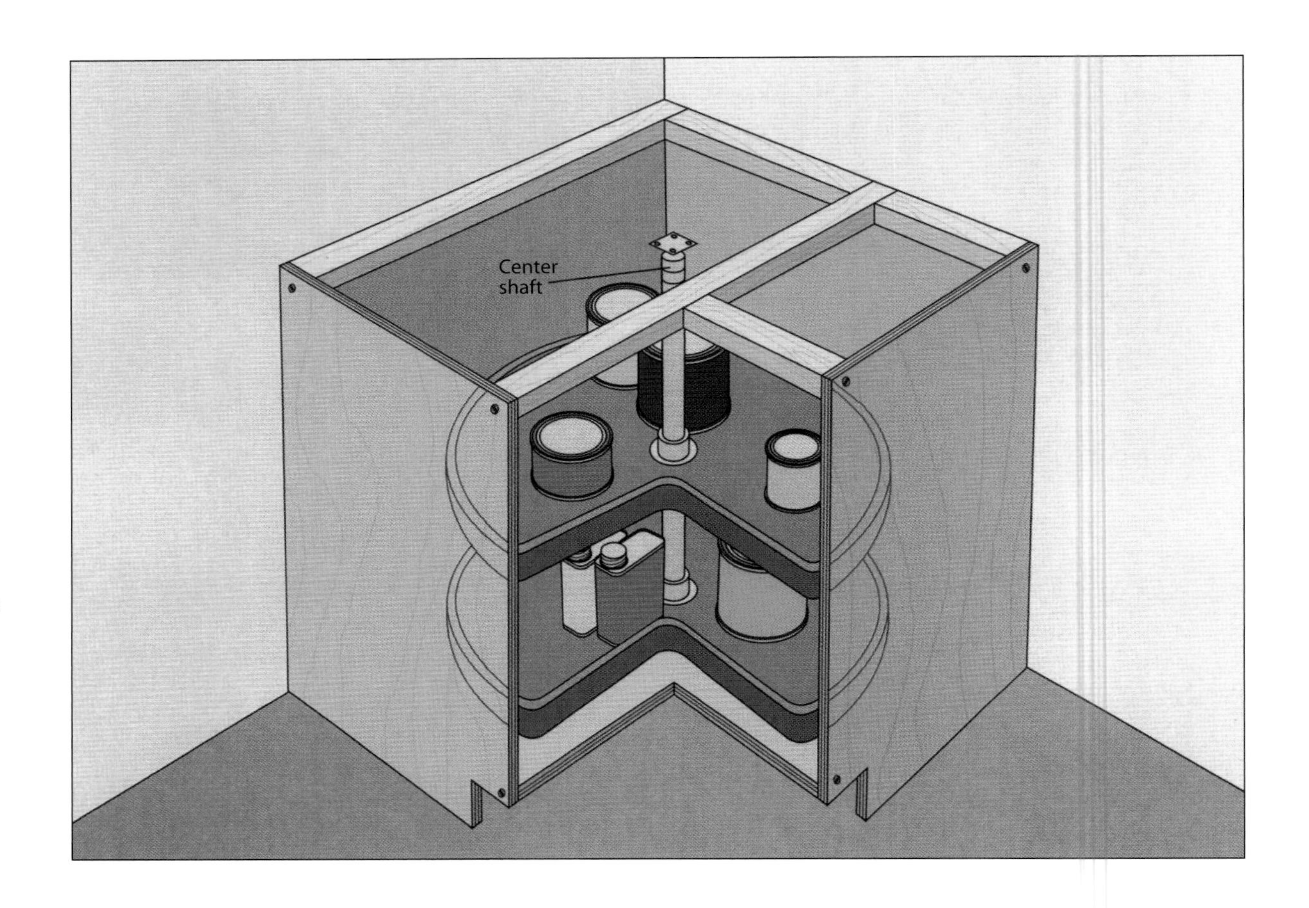

Using a Lazy Susan-type storage cupboard
If your workshop has an unused corner—an area under a counter, for example, install a commercial Lazy Susan-type cupboard to store workshop tools and supplies. The design of the device makes any item on the trays easily accessible. The model shown above features a carousel with two trays that revolve around a metal shaft. Using ¾-inch plywood, build a cabinet like the one shown above to house the carousel and support the metal shaft at both the top and bottom. Assemble the carousel following the manufacturer's instructions.

Shop Tip

A magnetic tool rack
Keep metal tools organized and accessible on a commercial magnetic tool rack. The model shown features a heavy-duty bar magnet that will hold any iron-based tool securely—from screwdrivers, chisels, and hammers to try squares and scissors. To mount the rack, screw the magnet to a wood strip and anchor the strip to wall studs above your workbench.

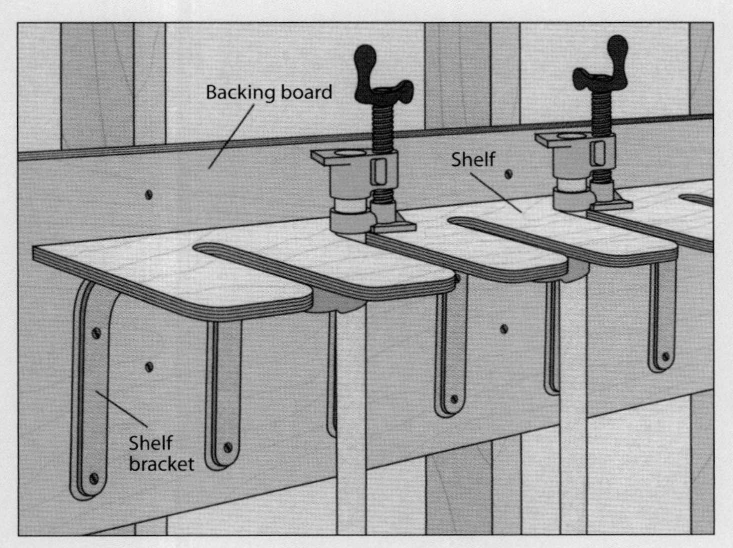

A Shelf for Clamps

Built from ½-inch plywood, the shelf shown at right features a series of notches for supporting bar and pipe clamps along a shop wall. Cut the shelf about 10 inches wide and as long as you need for the number of clamps you wish to store. Cut the notches at 3-inch intervals with a saber saw and make them wide enough for the clamp bars or pipes; 1¼ inches is about right for most clamps. Then screw shelf brackets to the underside of the shelf, centering them between the notches. Fasten the shelf to a backing board of ½-inch plywood, then anchor the board to the wall studs.

Shop Tip

Storing clamps in a can
A trash can fitted with a shop-made lid serves as a convenient way to store small bar or pipe clamps. Cut a piece of ½-inch plywood into a circle slightly smaller than the diameter of the can's rim. Then scribe a series of concentric circles on the plywood to help you locate the holes for the clamp bars. Space the circles about 3 inches apart and mark points every 3 inches along them. Bore a 1-inch-diameter hole through each point, fit the piece of plywood in the can and drop the clamps through the holes.

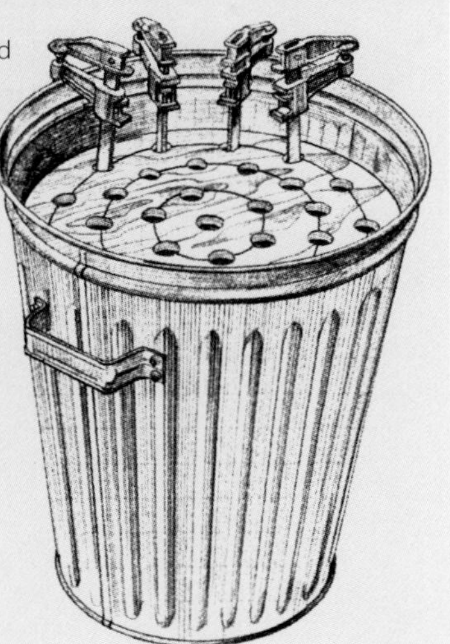

Mobile Clamp Rack

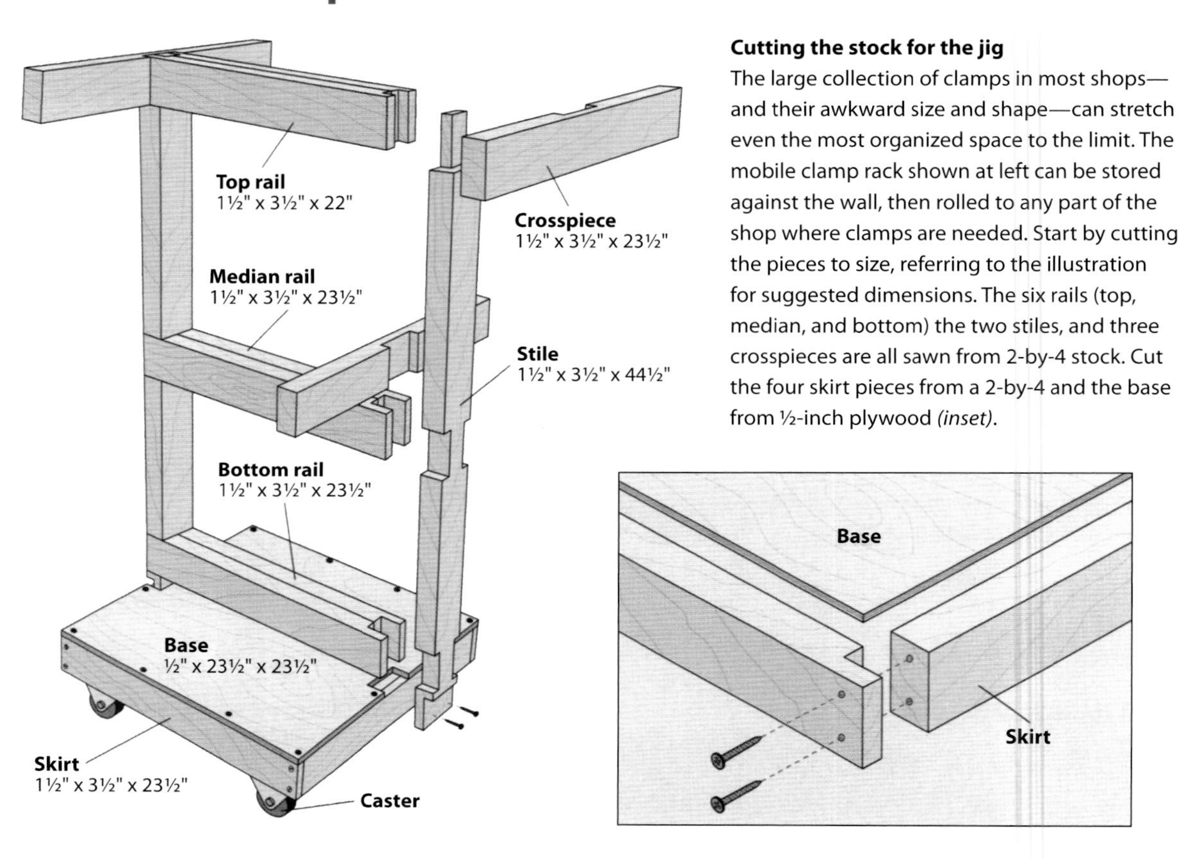

Cutting the stock for the jig

The large collection of clamps in most shops—and their awkward size and shape—can stretch even the most organized space to the limit. The mobile clamp rack shown at left can be stored against the wall, then rolled to any part of the shop where clamps are needed. Start by cutting the pieces to size, referring to the illustration for suggested dimensions. The six rails (top, median, and bottom) the two stiles, and three crosspieces are all sawn from 2-by-4 stock. Cut the four skirt pieces from a 2-by-4 and the base from ½-inch plywood *(inset)*.

Attaching the rails to the stiles

Prepare the rails for the joinery by cutting end rabbets that will fit into notches and dadoes in the stiles. The rabbets should be 1½ inches wide and ¾ inch deep, except for the top rails, which require a rabbet only 1 inch wide. Notch the top end of each stile on three sides, then rout back-to-back dadoes near the bottom end and middle of the stiles; make the dadoes 3½ inches wide and ¾ inch deep. Also cut a notch 3½ inches wide and ¾ inch deep from the bottom of each stile. When you assemble the rails and stiles, align the two halves of each rail face-to-face *(right)* and attach it to the stiles with screws.

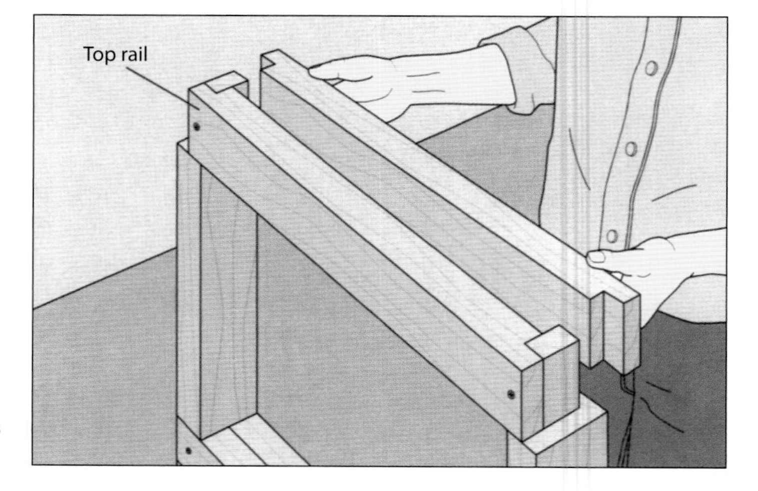

Mounting the crosspieces to the stiles

To join the crosspieces to the rack, cut a 3½-inch-wide dado in the middle of each piece and screw them in place *(right)*. The middle crosspiece will rest on the median rail. The top pieces will rest on the outside shoulders of the notched top of each stile.

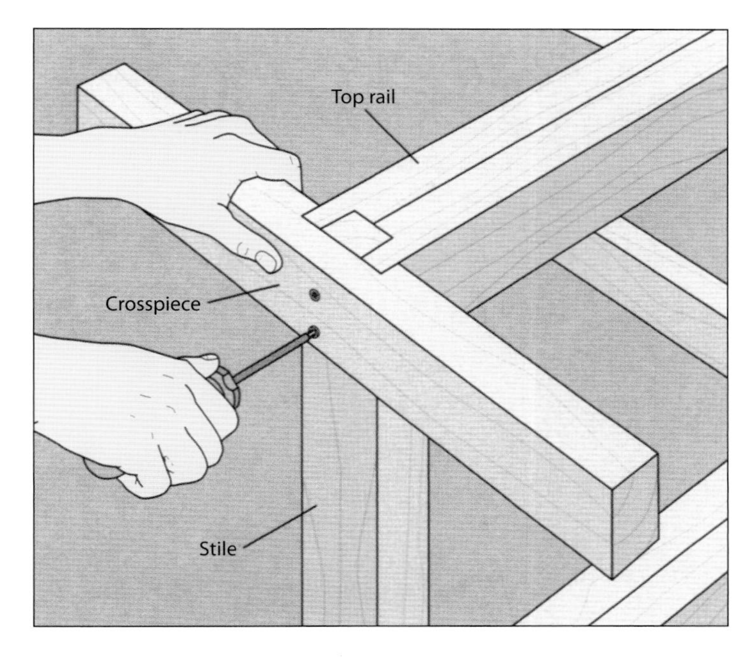

Attaching the stiles to the base

Finish the rack by sawing two notches in the base and skirt to accommodate the stiles, rabbeting one end of each skirt piece, and screwing them together to form a box *(page 91)*. Use screws to attach the base to the skirt. Finally, attach casters to the underside of the skirt at each corner of the rack, then slip the stiles into the notches in the base *(left)* and secure the stiles to the base and skirt with glue and screws.

Two Wall Racks for Clamps

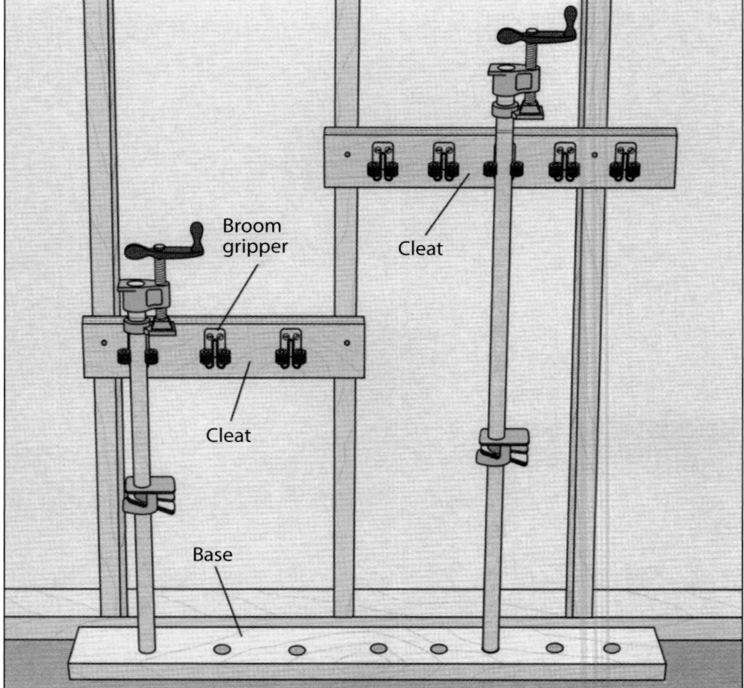

Making and installing the racks

Shop walls make ideal storage areas for bar and pipe clamps. For bar clamps *(above, left)*, nail two cleats across the wall studs. Position the upper cleat—made of plywood—high enough to keep the clamps off the floor; make the lower one from two 2-by-4s nailed together so that the clamps will tilt toward the wall. For pipe clamps *(above, right)*, nail cleats of ¾-inch plywood to the studs and screw broom grippers to the cleats. Position the cleats on the wall so the clamps will rest about 1 inch off the floor. Then cut the base from 1-by-4 stock, and bore a row of holes into it at the same interval as the grippers. Fasten the base along the floor so the holes line up with the clamps.

Shop Tip

Storing handscrews

For small clamps like handscrews, it is often unnecessary to build separate storage. An exposed stud can serve as an effective clamping post. Avoid storing clamps directly overhead, should vibration cause one to fall. If your shop has a finished ceiling and walls, install a length of 2-by-4 across the legs of your bench and secure your clamps to the stock.

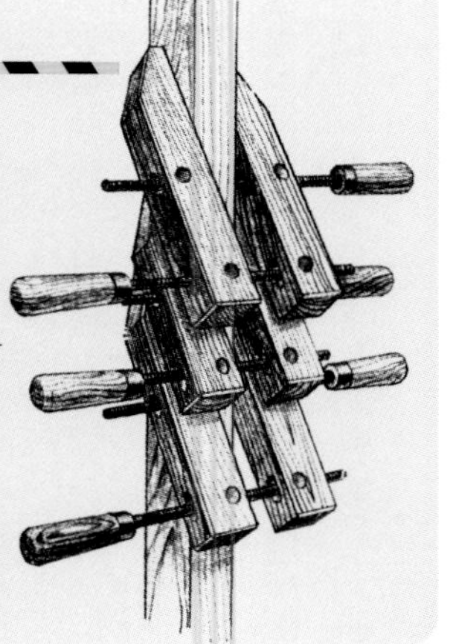

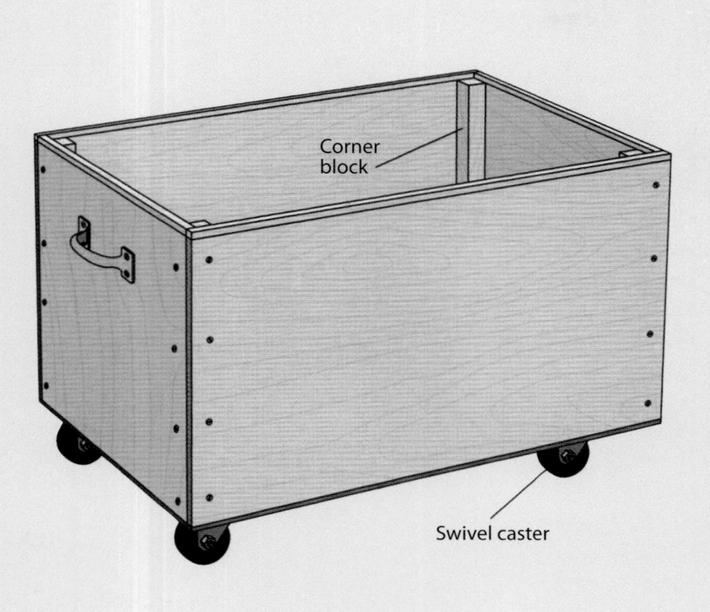

A Scrapbox

Use a scrapbox to keep from cluttering the shop floor with cut-offs, shavings, and other refuse. The design shown at left can be built quickly from ¾-inch plywood; casters allow the unit to be rolled where it is needed and moved out of the way when it is not. Saw the sides and bottom to a size appropriate to your needs, then cut four corner blocks from 2-by-2 stock. Screw the four sides together, driving the screws into the corner blocks. Turn the box over and nail the bottom to the corner blocks and sides. Add a lip around the top to hide the plywood edges. Finally, screw casters to each bottom corner and a pull handle at one end.

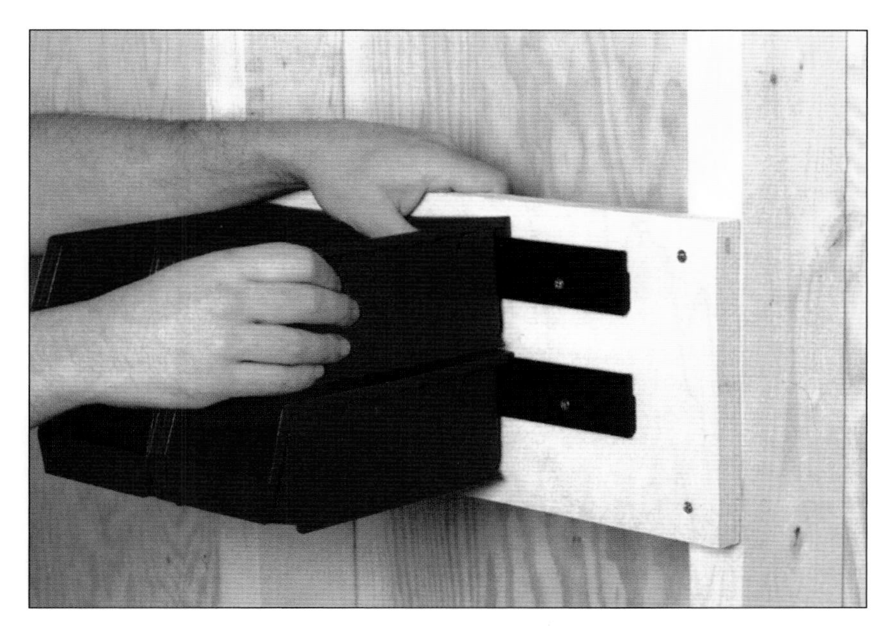

Some storage devices, particularly those designed for small items, are less trouble to buy than to build. The system shown at left features open plastic bins that can be lined up or stacked. The bins are suspended from plastic strips that are screwed to the wall.

Work Surfaces

It is a truism that no workshop is ever large enough; it is equally true that no woodworker ever has enough tables, benches, sawhorses, stands, or props to support work in progress. The traditional workbench, however useful or necessary *(see page 30)*, is only the beginning. For many uses, it is too high, too small, or too immobile to be helpful.

When it is time to mark the elements of a joint or assemble the many pieces of a chair, a solid work table, like the library-style table shown on page 95, can serve as the command center of your shop, becoming the focus of many operations. This design features a spacious work surface and sturdy construction. The only drawback is size: one would need a fairly large shop to accommodate this table. For a smaller shop with cramped quarters, consider the folding table featured on page 97. Offering almost as much surface area as the library-style version, it can be folded out of the way against the wall when it is not needed. The temporary table illustrated on page 99 offers yet another solution to the constant conflict between space and convenience, satisfying both the need for a substantial working surface and ease of storage. Resting on sawhorses, this plywood sheet tabletop can be set up whenever a flat surface is required, then be dismantled and put away when your project moves on to another phase.

With its myriad uses, the sawhorse is also the workhorse of the shop. A sawhorse can serve as a set of legs for a fold-down work table *(page 98)* or a simple prop for sawing stock. With a few notches cut into their crosspieces, horses can form part of the frame for a shop-made glue rack *(page 106)*. Clamped to a 2-by-4 attached to a commercial roller, a sawhorse becomes a custom-built roller stand.

Outfeed tables and roller stands that hold unwieldy panels or long planks significantly expand the versatility of tools like table saws, band saws, and drill presses. Set up at the same level as a machine's table, or fractionally below it, these props can be as welcome as a second set of hands, enhancing a tool's capacity to handle large workpieces efficiently and safely.

Work surfaces can even be rigged to compensate for a lack of full-size stationary machines. The stand shown on page 116 is designed to let you mount a benchtop tool at a comfortable working height. The three-in-one tool table featured on page 118 can transform a router, saber saw, and electric drill into mini-stationary tools.

Given a need and a few pieces of wood, every woodworker will devise some way to improve his or her tools. The examples that follow are mere suggestions, for it is impossible to limit the imagination when the need arises for improving the workshop.

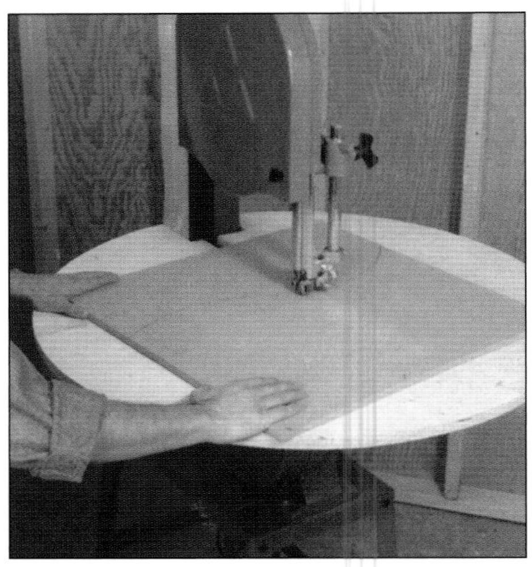

Better control produces better results. Secured to a band saw's original table, a shop-made extension table keeps a large hardboard sheet level during a curved cut.

The door of this storage cabinet folds down from a shop wall to become a sturdy work surface. Supported by solid lumber legs, it is an ideal work table for light-duty operations such as gluing up and assembling small carcases. For details on how to built this unit, see page 84.

MARPLES

Work Tables

For many light woodworking chores, from marking out joints to assembling pieces of furniture, a simple work table fits the bill as well as a traditional woodworker's bench. This section features several table designs. All are quick, easy, and inexpensive to build. The table shown opposite is sufficiently large and sturdy for most jobs; if space is at a premium, a good compromise would be one of the fold-up versions shown on pages 97 and 98. You can also conserve space by incorporating storage shelves, drawers, or cabinets in your design. For assembling carcases and other pieces of furniture, you may find the low-to-the-ground table on page 96 handier than a standard-height work surface.

Whichever design you choose, be careful of the nails or screws you use to construct a table—particularly when fastening the tabletop to the frame. Take the time to countersink or counterbore screw heads and set nail heads below the surface to prevent the fasteners from marring your work.

Almost as strong as a traditional workbench, this commercial work table is a versatile workhorse, especially when paired with a woodworker's vise. The cabinet and drawers provide storage space, and can be locked to secure valuable tools.

Despite its lightweight, compact design, the Black & Decker Workmate™ can support loads up to 550 pounds. It also folds virtually flat for easy storage. A special pivot design allows the vise jaws to be angled, for securing workpieces like the tapered leg shown in the photo. This particular Workmate™ features a storage tray and a top that flips up for vertical clamping. The Workmate™ has a long, colorful history. By 1968, the prototype, featuring a patented folding H-frame, had been rejected by every major tool manufacturer in Britain. Four years later, the inventor of the Workmate™, Ron Hickman, persuaded Black & Decker in England to mass produce his invention. International distribution rights were negotiated the following year. Popular success for the Workmate™ was almost immediate.

A Library-Type Work Table

The all-purpose table shown below is built with a combination of lumber and plywood. Refer to the dimensions in the illustration for a work surface that is 5 feet long, 3 feet wide, and 3 feet high.

Saw the legs to length from 4-by-4 stock, then prepare them for the rails: Cut a two-shouldered tenon at the top end of each leg with shoulders ¾ inch wide *(inset)*. Next, cut the rails, stretchers, and braces to length from 2-by-4s. Saw miters at both ends of the braces so that one end sits flush against the inside edge of the legs and the other end butts against the bottom of the rails. Prepare the front, back, and side rails for assembly by beveling their ends and cutting rabbets to accommodate the leg tenons *(inset)*. Screw the stretchers to the rails, spread glue on the contacting surfaces of the legs and rails, fit the pieces together, and screw the rails to the legs. Next, attach the braces to the legs and rails with screws.

Cut the tabletop from ¾-inch plywood and screw it to the rails. Finally, cut a piece of ¼-inch hardboard to the same dimensions as the top and nail it to the plywood as a replaceable protective cover. Be sure to set the nail heads below the surface.

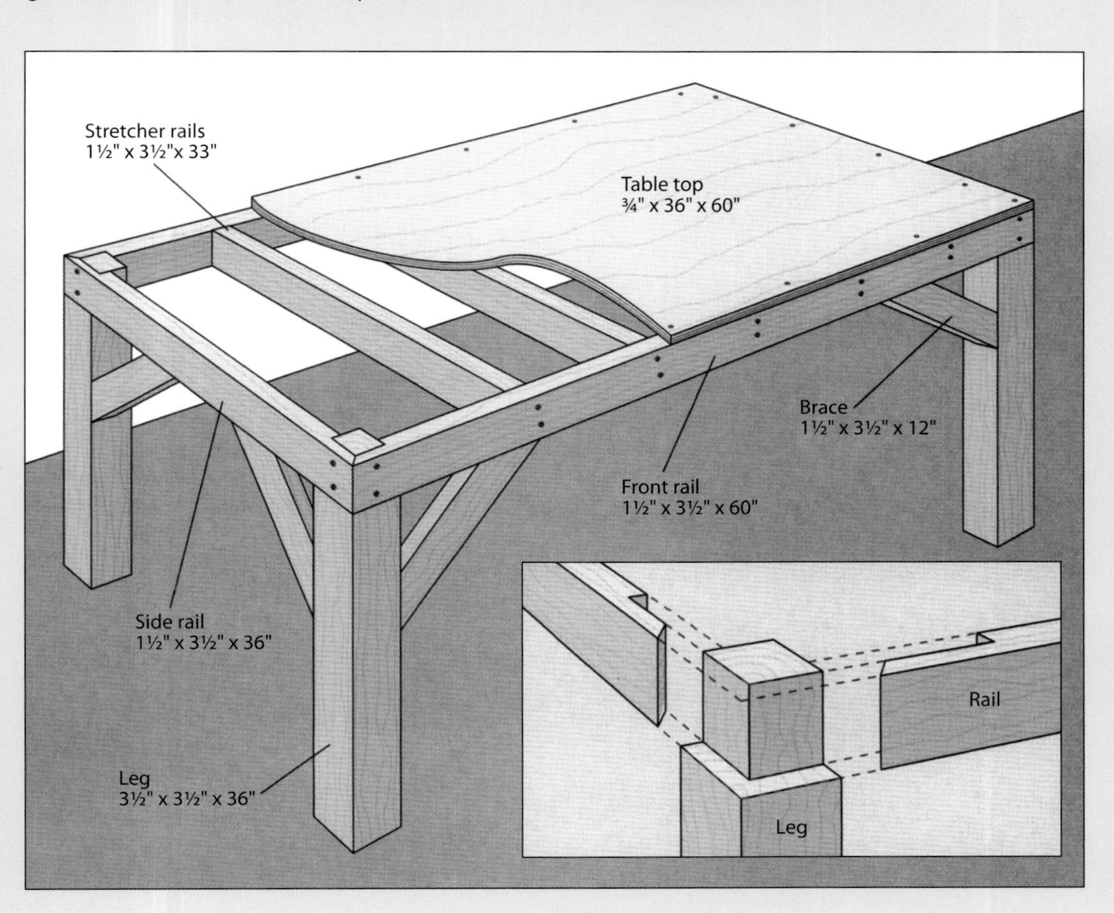

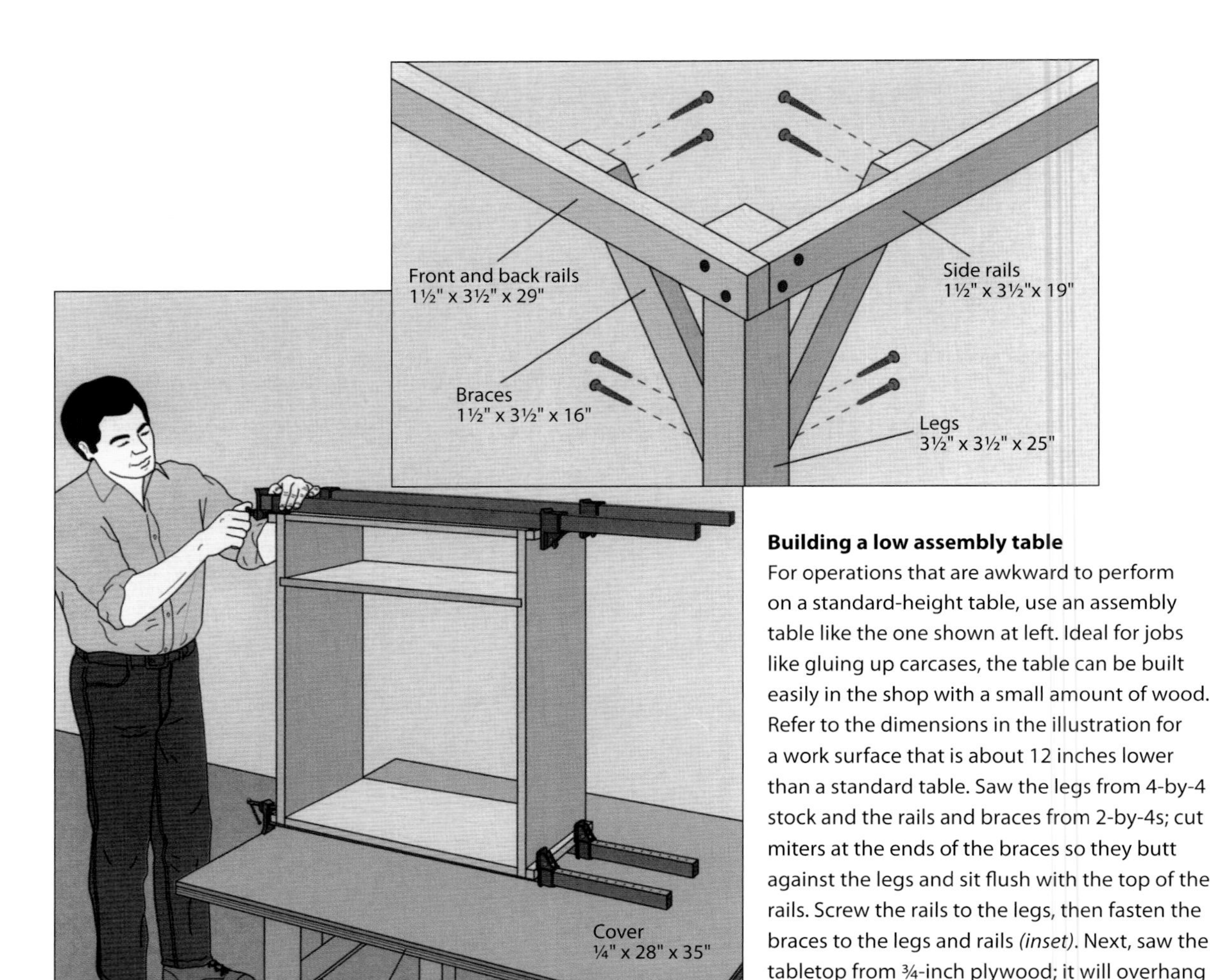

Building a low assembly table

For operations that are awkward to perform on a standard-height table, use an assembly table like the one shown at left. Ideal for jobs like gluing up carcases, the table can be built easily in the shop with a small amount of wood. Refer to the dimensions in the illustration for a work surface that is about 12 inches lower than a standard table. Saw the legs from 4-by-4 stock and the rails and braces from 2-by-4s; cut miters at the ends of the braces so they butt against the legs and sit flush with the top of the rails. Screw the rails to the legs, then fasten the braces to the legs and rails *(inset)*. Next, saw the tabletop from ¾-inch plywood; it will overhang the rails by about 3 inches on all sides. Screw the top to the rails, countersinking the fasteners. Cut a replaceable cover from ¼-inch hardboard and nail it to the tabletop; set the nail heads below the surface of the cover.

Stow-away Work Tables

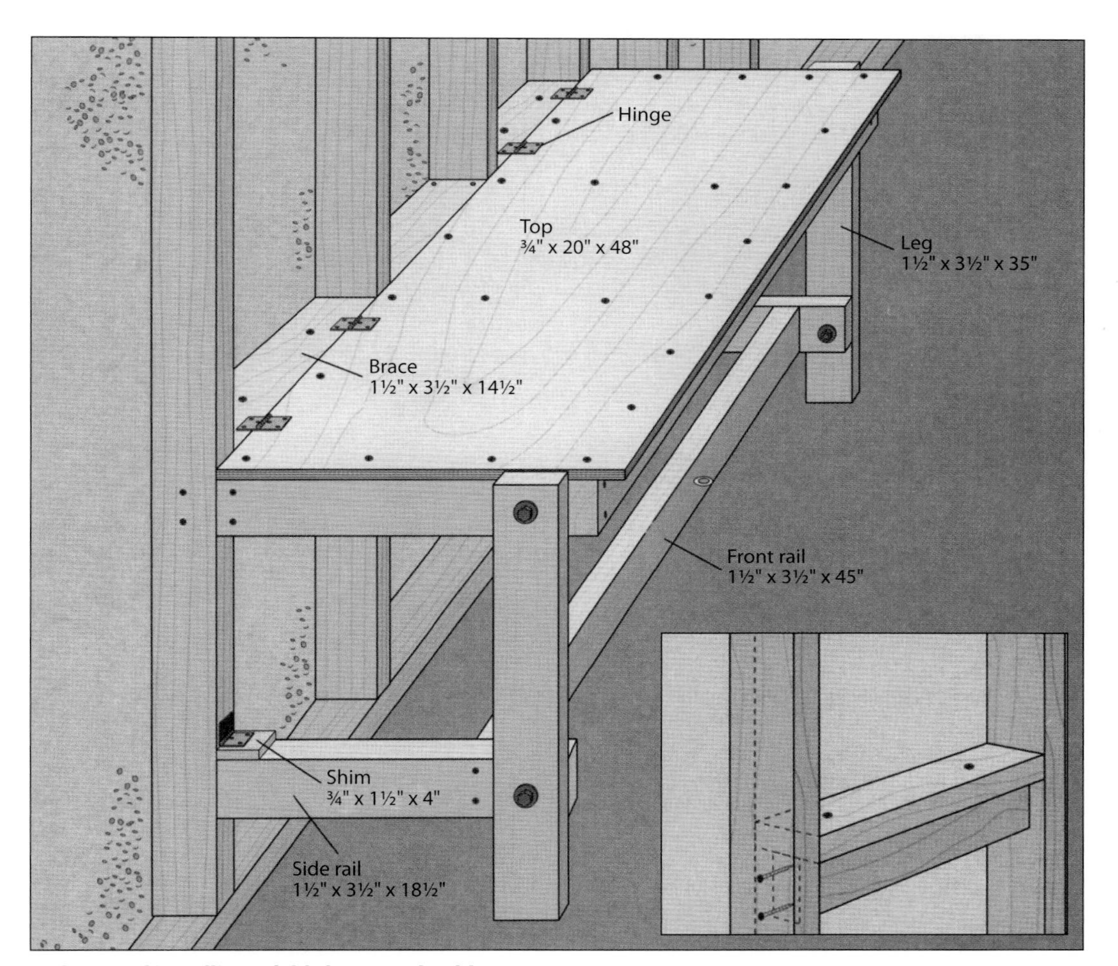

Making and installing a fold-down work table
The table shown above incorporates a large and sturdy work surface, but still conserves space by folding up against a wall when it is not in use. The dimensions in the illustration yield a work surface measuring 20 by 48 inches. Cut the bracing, legs, rails, and stretchers from 2-by-4 stock and screw the bracing between the wall studs *(inset)*; there should be one brace for every pair of studs along the table's length. Fasten the front legs to the side rails using carriage bolts and lock nuts; place washers on both sides of the legs. Leave the bolts just loose enough to allow the legs to pivot when the table is folded up. To complete the frame, attach the front rails to the side rails. Add a 45-inch-long top rear rail and fasten two 17-inch-long stretchers between the rear rail and the top front rail to provide added support for the top. Next, screw the top to the rails, countersinking the fasteners. To allow the table to fold down without binding, screw shims to the ends of the bottom side rails, then attach the table to the bracing with butt hinges; use two hinges for each outside brace. Finally, drive an eye bolt into the bottom front rail and a catch into the wall to secure the table when it is folded up.

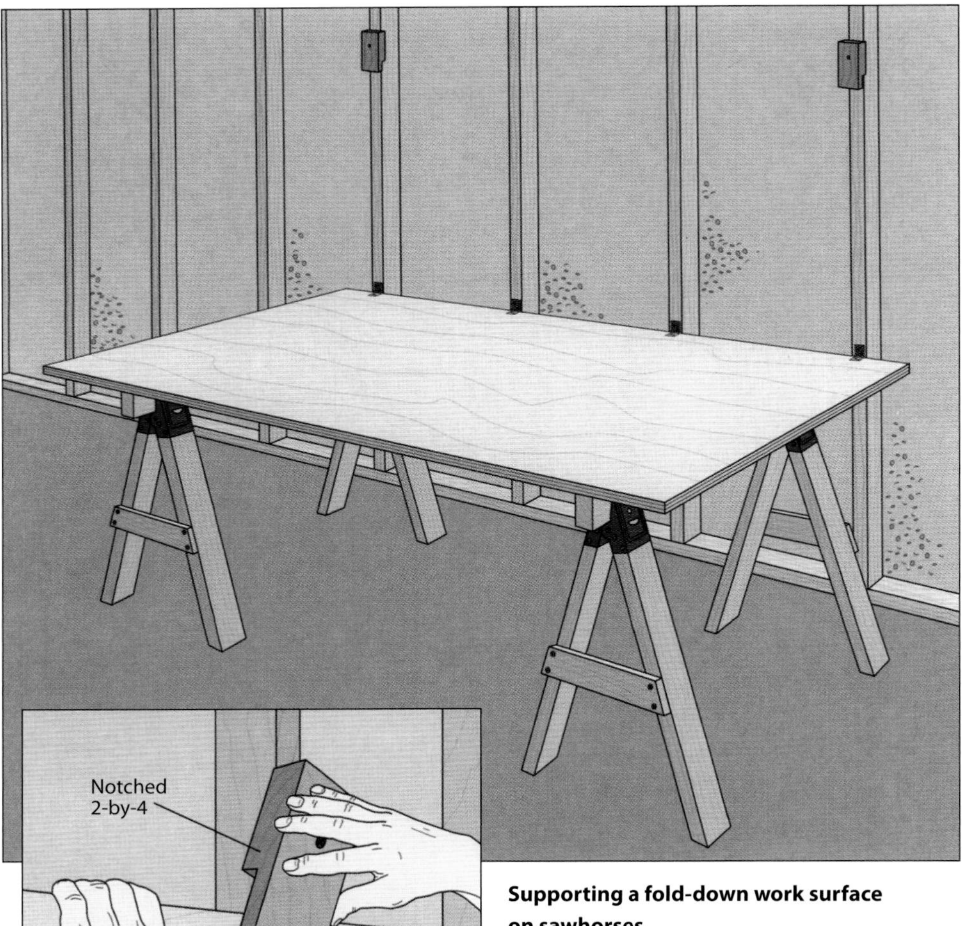

Supporting a fold-down work surface on sawhorses

Rather than building a framework for a fold-down work surface, you can use a panel of ¾-inch plywood hinged to the wall and supported by sawhorses. The surface can be of any size. Begin by setting the panel on two sawhorses; one edge of the panel should be flush against the wall. Mark a point on the panel at every wall stud, then install butt hinges, screwing one leaf of each hinge to a stud and the other leaf to the panel at a pencil mark. To secure the panel when it is folded up, screw a notched piece of 2-by-4 to the stud closest to the middle of the panel at a height that will allow the notched end to slip over the edge of the panel *(inset)*.

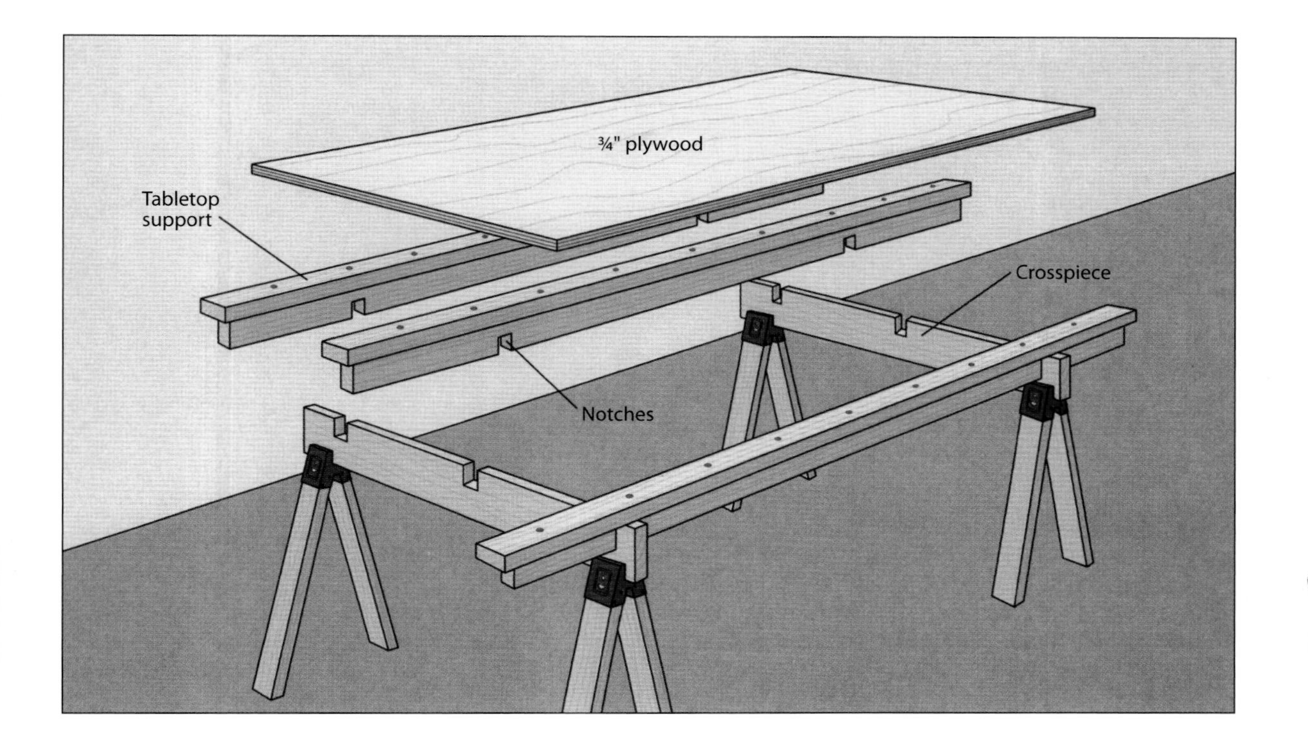

Setting up a temporary work surface
Consisting of two sawhorses, six 2-by-4s and a plywood panel, the unit shown above is inexpensive and easy to put together, yet it provides a large and stable work surface that can be set up and disassembled quickly. Start by fitting the sawhorses with crosspieces cut from 2-by-6 stock, then cut the 2-by-4s to the same length as the panel. In three of the boards, cut a notch about 8 inches from each end; the notches should be about 2 inches deep and as wide as the thickness of the crosspieces. Cut matching notches in the top edges of the crosspieces. Center the unnotched edge of the notched boards along the face of the other 2-by-4s and screw them together to form three T-shaped tabletop supports. The sawhorse supports can be used to hold a large sheet of plywood for ripping, or a permanent top can be screwed to the 2-by-4s.

Shop Tip

Stabilizing a temporary work surface
Tables consisting of plywood panels laid across two sawhorses are a cinch to set up, but they tend to slide and twist. Attaching cleats to the underside of the panel on either side of the sawhorses' crosspieces helps to stabilize these makeshift work surfaces. To position the cleats, mark the outline of each crosspiece on the panel, then screw a length of 1-by-3 to the panel on either side of each outline, leaving as little clearance as possible between the cleats and the crosspieces. For extra stability, attach 1-by-3s between the pairs of cleats to serve as rails.

Sawhorses

Sawhorses have countless uses in the woodworking shop, from table legs to tool stands. Occasionally it seems that their original purpose—to support boards for sawing—is only an afterthought. It is easy to see why sawhorses are considered so versatile, for their compact design makes them especially useful in shops with limited floor space. Some commercial models, like the ones in the photo at right, can be adjusted to different heights and folded up for easy storage. With commercial brackets *(right)*, you can size sawhorses to suit your needs. The shop-made horses featured on page 102 can be disassembled and put away after use.

Different operations require different-sized sawhorses. For supporting stock for handsawing or holding large workpieces at a comfortable height, small horses about 18 inches high are ideal. Taller sawhorses are needed if they are to be used to hold up a work surface or as outfeed supports for a table saw. They should be about ¼ inch lower than the saw table.

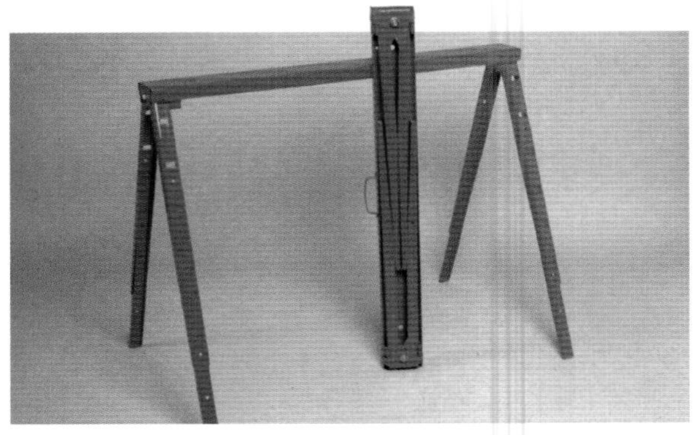

This sawhorse features leg extensions that can he adjusted to a variety of heights. The legs retract into the crosspiece, making the unit compact and portable. A pair of these slender metal horses can support one ton of material.

Whatever the dimensions of your sawhorses, never make them taller than their length, as they will tend to be unstable.

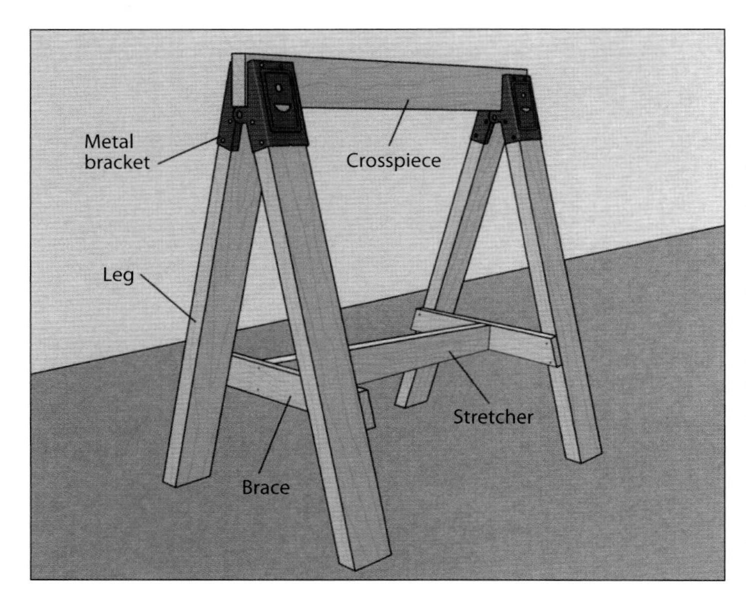

Making Sawhorses

Using commercial sawhorse brackets

A pair of metal sawhorse brackets can help you transform a couple of 2-by-4s and 1-by-3s into a sturdy sawhorse, like the one shown at right. Saw the legs and crosspiece from 2-by-4s, then cut a bevel at the bottom of the legs so they will sit flat on the floor. Fit the legs into the bottom of the brackets, insert the crosspiece and spread the legs; the brackets will grip the crosspiece and stabilize the horse. Screw the brackets to the legs and crosspiece. For added stability, add braces and a stretcher. The braces are cut from 1-by-3s and screwed to the legs, making sure that the ends are flush. For the stretcher, cut a 1-by-3 to size and screw it between the braces.

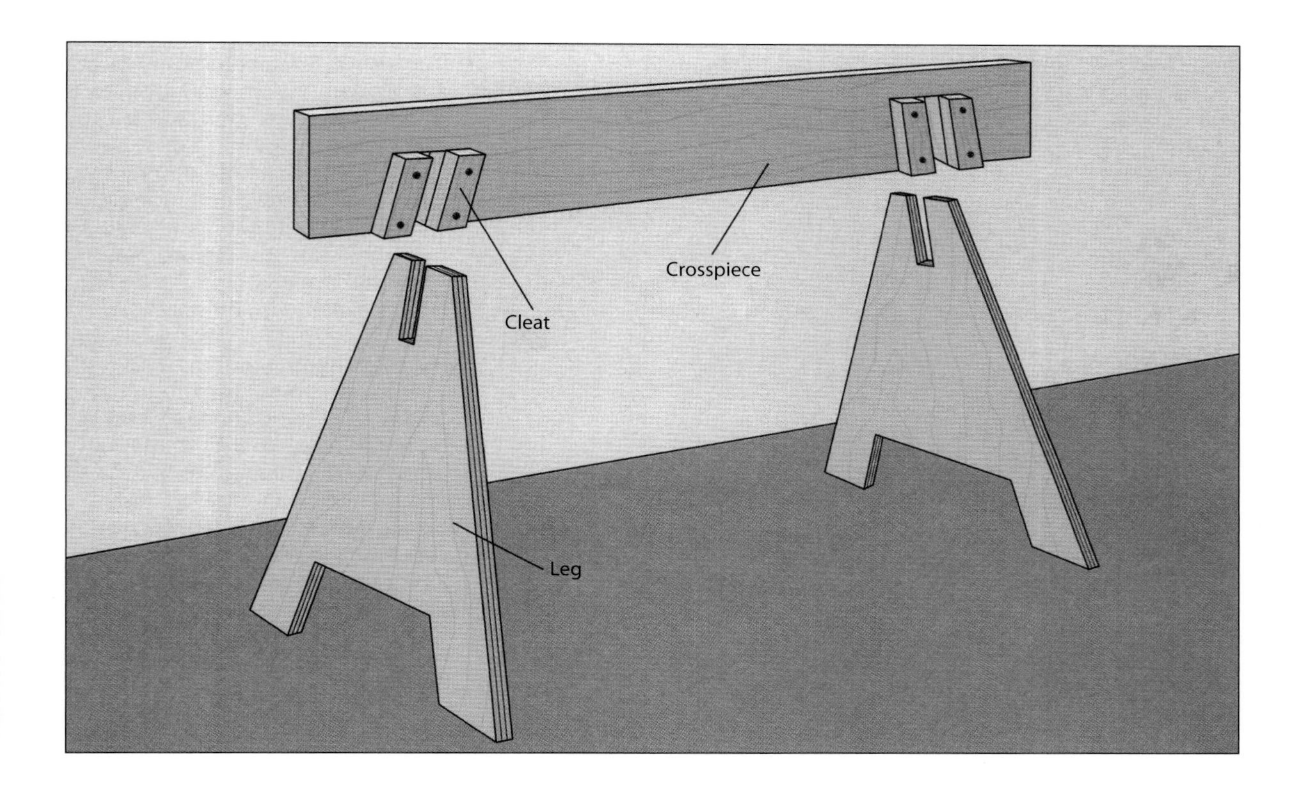

Building a knock-down sawhorse
With only a small amount of lumber and plywood and a few minutes' time, you can make a sturdy, knock-down sawhorse like the one shown above. Cut the legs from ¾-inch plywood, then saw a 3-inch-deep notch in the middle of the top of both pieces. Next, cut the crosspiece from 1-by-6 stock and saw a 1½-inch-deep slot 8 inches in from either end to fit into the legs. Angle the slots roughly 5° from the vertical so the legs spread slightly outward. For added stability, screw 4-inch-long 1-by-2 cleats to the crosspiece on each side of the slots.

Shop Tip

Stacking sawhorses
If you want to stack your sawhorses, instead of the type of braces seen on page 100, make plywood braces like those shown at right. Cut a notch in each brace so it will mesh with the crosspiece of the sawhorse beneath it. To prevent the stack from toppling over, be sure the fit is snug by leaving only a small amount of clearance in the notches.

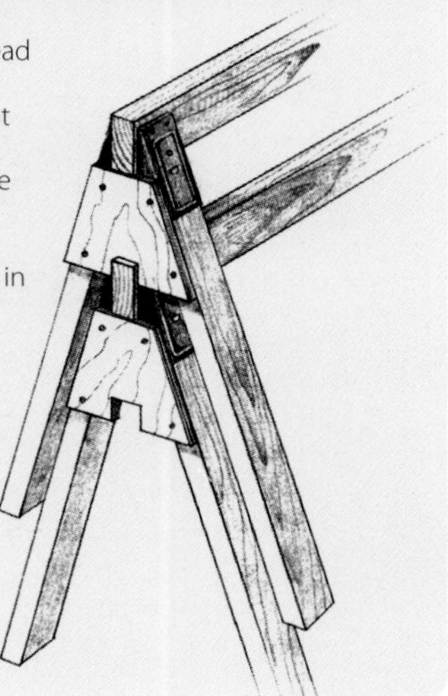

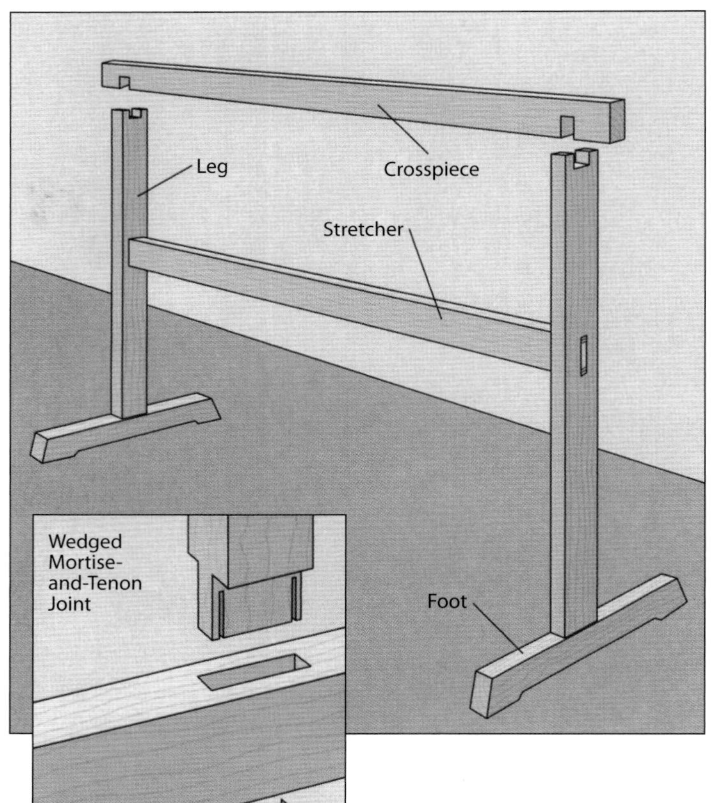

Assembling a frame-and-foot sawhorse
Lightweight, compact frame-and-foot sawhorses like the one shown at left can be built from 2-by-4 stock. Start by cutting the legs to a suitable height, then prepare them to join to the other parts of the unit: Cut tenons at the bottom ends, rout through mortises halfway up the faces, and saw 1-inch-deep notches in the middle of the top ends. Cut the feet to length and, for added stability, cut recesses along their bottom edges, leaving a 2-inch pad at each end. Rout mortises through the middle of the feet for the leg tenons. Next, saw the stretcher to fit between the legs and cut tenons at both ends. Cut the crosspiece and saw a notch 4 inches from either end that will fit into the notch at the top of the legs. To reinforce the mortise-and-tenon joints, saw a pair of kerfs in the end of each tenon and make wedges to fit into the kerfs *(inset)*. Tap the wedges in to expand the tenon when the joint is assembled.

Shop Tip

Padding sawhorses
To prevent a sawhorse from marring your work, cover its crosspiece with a strip of old carpet. Fold the carpet over the top edge of the crosspiece and screw it to the sides. For a smoother surface, use an old towel or blanket rather than a piece of carpet.

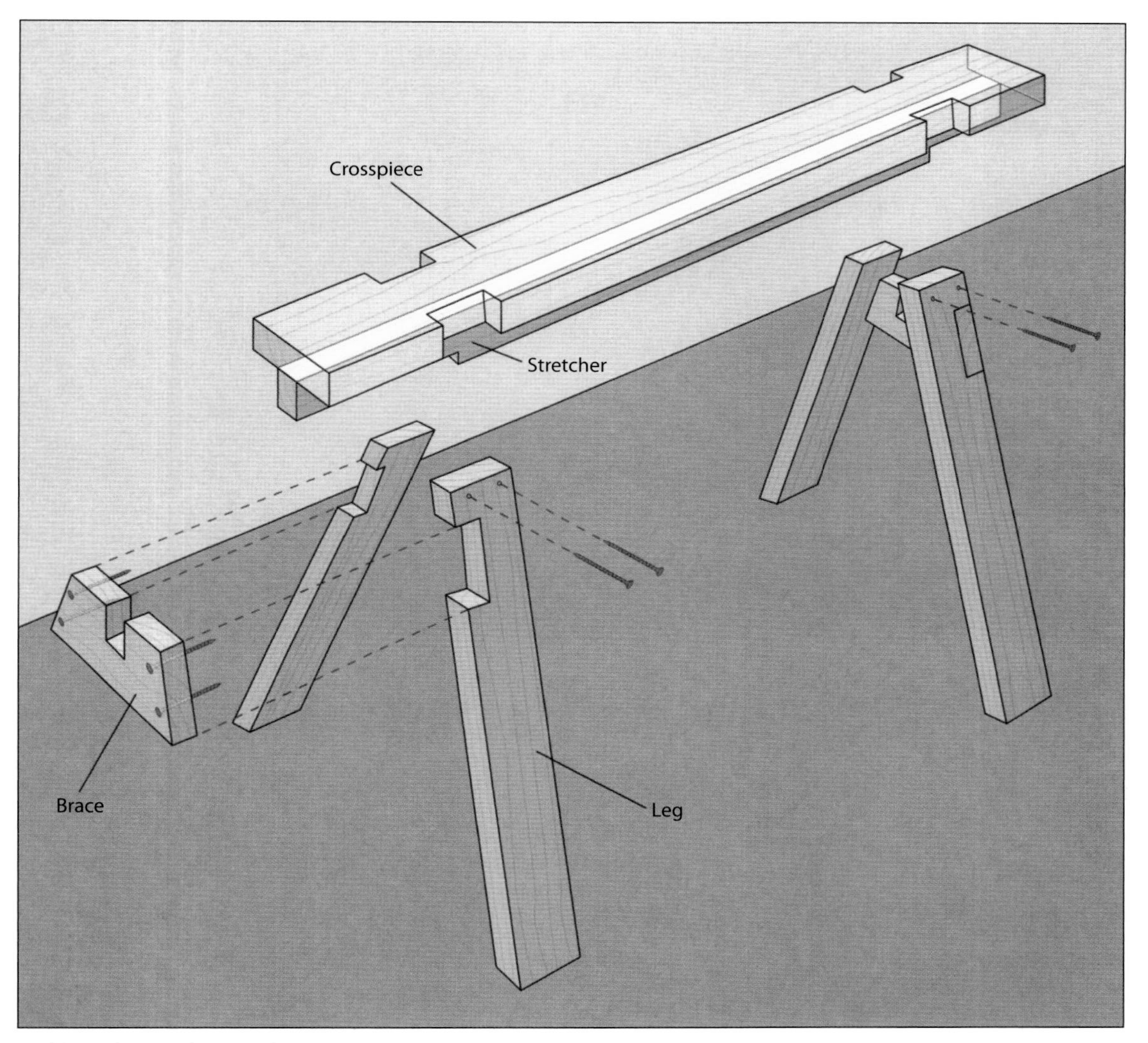

Making a heavy-duty sawhorse

Reinforced by a stretcher, braces, and simple joinery, the sawhorse shown above will endure for years as a sturdy work surface. Saw the crosspiece to length from a 2-by-6 and cut dadoes in the edges about 4 inches from either end to accommodate the legs. Angle the dadoes roughly 10° from the vertical. Next, saw the 2-by-4 legs to length and cut 1½-inch-deep angled notches into their outside edges to house the braces. The top of each brace should rest about 1½ inches below the tops of the legs. Also cut bevels at both ends of the legs so they will sit flat on the floor and lie flush with the crosspiece. The stretcher is a 2-by-4 cut to the same length as the crosspiece; cut a notch in each end to line up with the brace, leaving a 1½-inch shoulder. Saw the braces from 2-by-6 stock, mitering the ends to be flush with the outside faces of the legs and sawing a 2-inch deep notch in the middle of the top edge for the stretcher. Finally, glue up the sawhorse, strengthening the joints between the legs, crosspieces, and braces with screws.

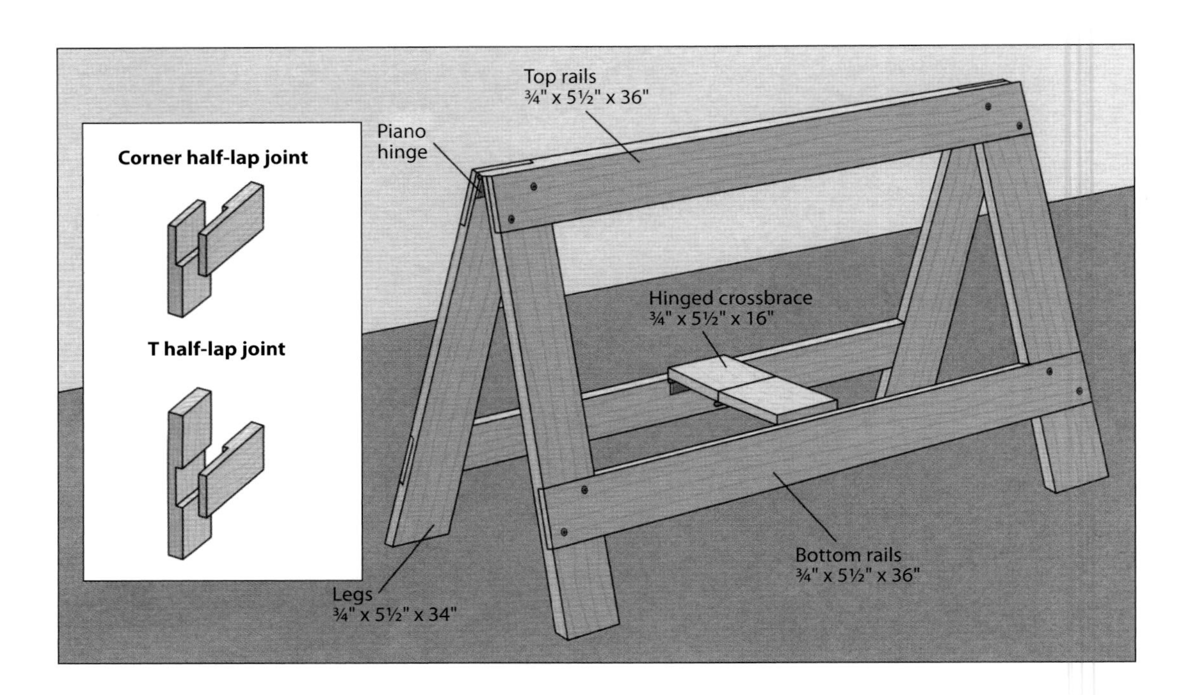

Building a folding sawhorse
Made entirely from 1-by-6 stock, with a hinged crossbrace and top, this light-weight sawhorse folds flat to store easily in even the most cramped workshop. Cut the legs and rails to length. Then, cut notches in the pieces for half-lap joints. Use T-type half-laps *(inset, bottom)* to join the legs to the bottom rails, and corner half-laps *(inset, top)* to join the top rails to the legs. Assemble and glue the two sections of the horse, and reinforce the joints with screws. When the glue has cured, join the two sections at the top rails with a continuous piano hinge. Finally, cut the crossbrace; be sure it is long enough so when the horse legs spread, the piano hinge is recessed between the top rails. Saw the crossbrace in half and connect the pieces with a piano hinge, making sure that the hinge is installed so the brace will pivot upwards. Then, fasten the crossbrace to both side rails, again using piano hinges.

Shop Tip

Securing workpieces edge-up on sawhorses
Clamp handscrews to the crosspieces of two sawhorses to support work edge-up when a bench vise is not available. To prevent the handscrews from pivoting, secure each with two C clamps as shown. Use as many sawhorses and handscrews as needed to adequately support the piece.

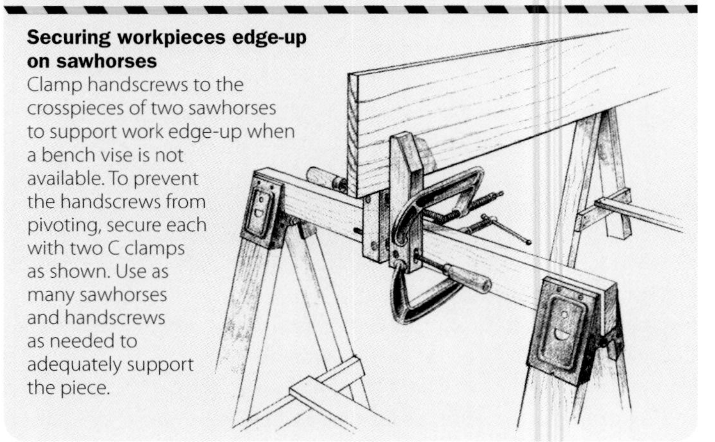

A Variable-Height Work Surface

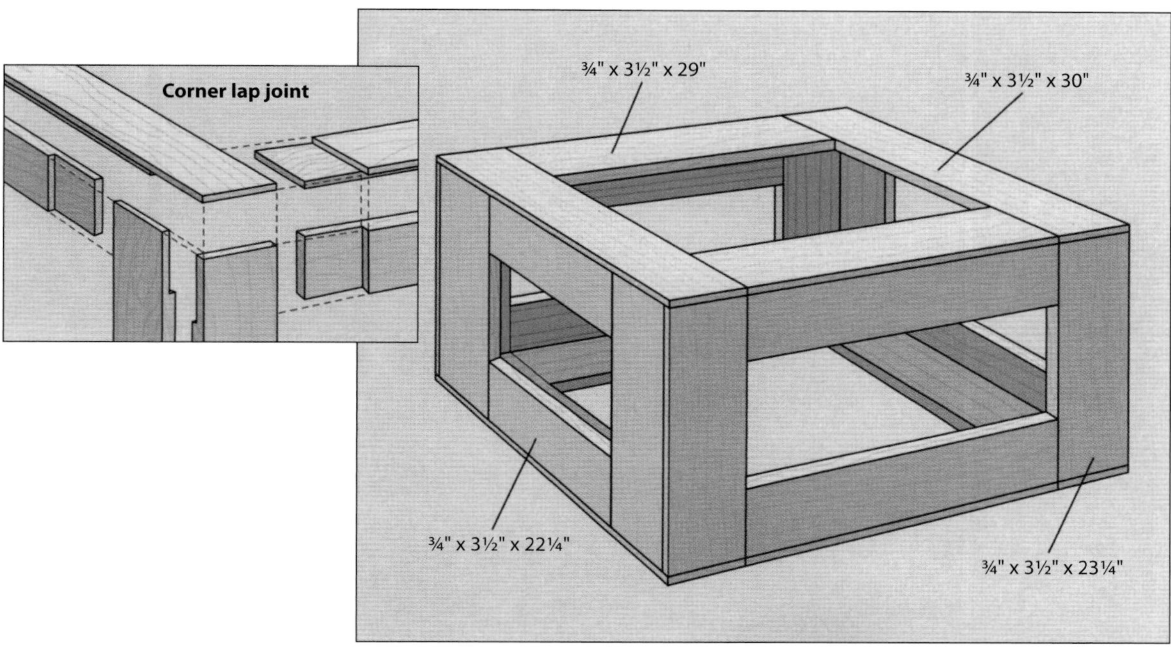

Building the box

Constructing a box with different width, length, and height dimensions will provide you with a work surface that can be used at three levels. The top surface of the box shown at right, for example, can be either 24, 30, or 36 inches high. Saw all the pieces from 1-by-4 stock, making four boards 30 inches long, eight 29 inches long, another four 22¼ inches long and eight more 23¼ inches long. Use a lightweight wood like pine to make the box easily portable. Assemble the box in two steps, gluing up the six sides individually, then joining them together. To make each side, cut rabbets at both ends of each board and glue up the boards with a lap joint *(inset)*. Once the glue has cured, cut a rabbet along the outside edges of all the sides, and glue up the sides into a carcase. If you choose to reinforce the joinery at the corners with screws, be sure to countersink the screw heads.

Edge-Gluing Jigs

Glue Rack

Building the jig

A pair of racks made from two saw-horses like the one shown at right provides a convenient way to hold bar clamps for gluing up a panel. Remove the crosspiece from your sawhorses and cut replacements the same width and thickness as the originals, making them at least as long as the boards you will be assembling. Cut notches along one edge of each cross-piece at 6-inch intervals, making them wide enough to hold a bar clamp snugly and deep enough to hold the bar level with the top of the crosspiece. You can also cut notches to accommodate pipe clamps, but it is better to use bar clamps with this jig since they will not rotate.

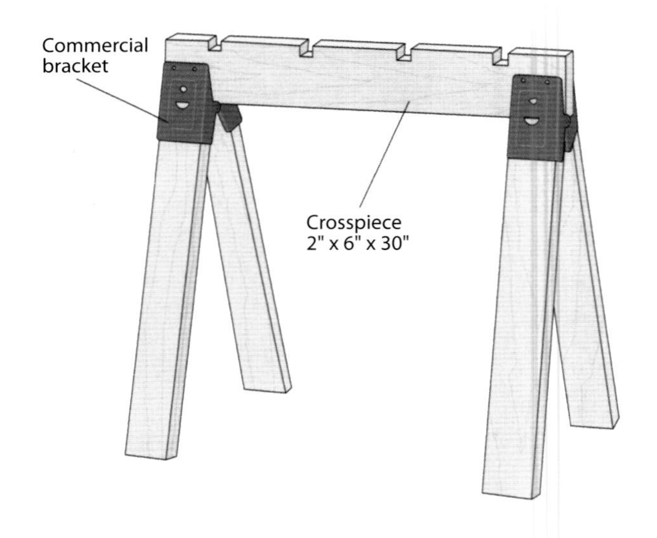

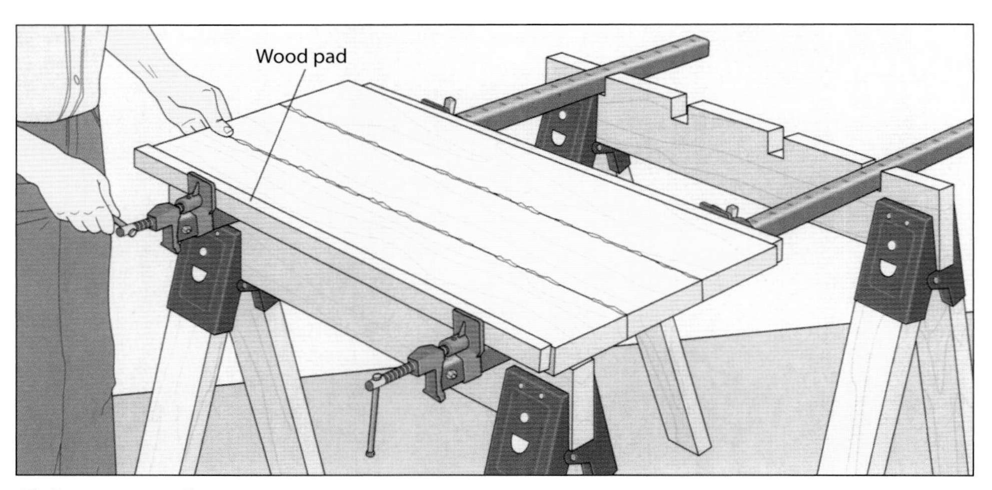

Gluing up a panel

Seat at least two bar clamps in the notches so that the boards to be glued are supported every 24 to 36 inches. To avoid marring the edges of the panel when you tighten the clamps, use two wood pads that extend the full length of the boards. Set the boards face-down on the clamps and align their ends. Tighten the clamps just enough to butt the boards together *(above)*, then place a third clamp across the top of the boards, centering it between the others. Finish tightening all the clamps until there are no gaps between the boards and a thin bead of adhesive squeezes from the joints.

Crossbars for Edge Gluing

Building the crossbars

To keep panels from bowing during glue-up when clamping pressure is applied, bolt a pair of crossbars like the one shown at right between each pair of clamps. Make each crossbar from two short wood spacers and two strips of 1-by-1 hardwood stock a few inches longer than the panel's width. The spacers should be slightly thicker than the diameter of the bolts used to hold the crossbars in place. Glue the spacers between the ends of the strips, and spread wax on the crossbars to prevent excess glue from adhering to them.

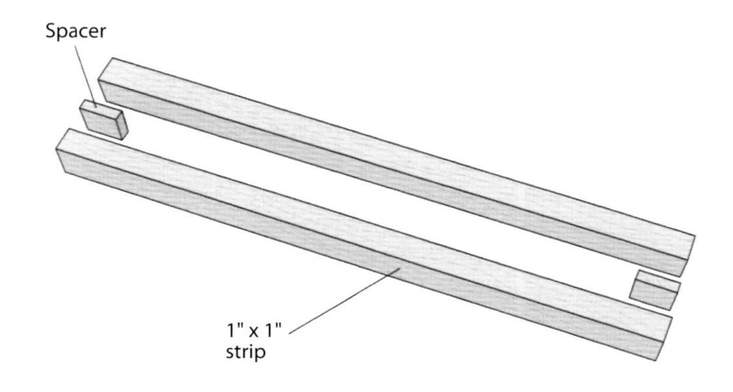

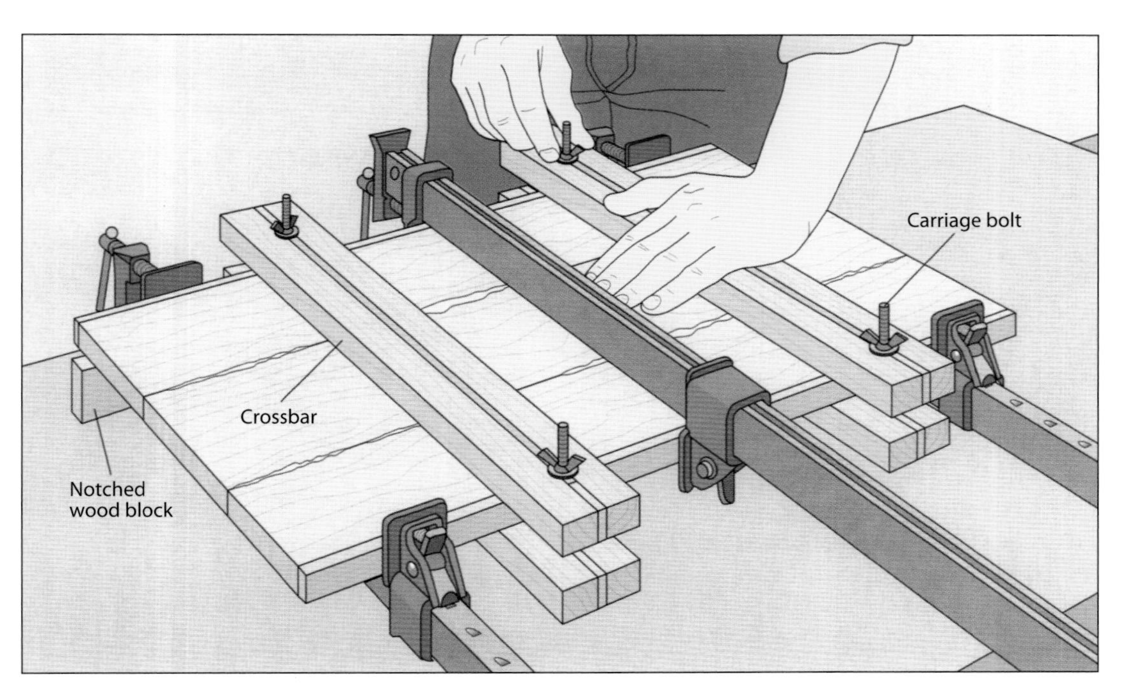

Installing the crossbars

Glue up the boards as you would on a rack *(page 106)*. To prevent the bar clamps from tipping over, place the end of each one in a notched block of wood. Before the bar clamps have been fully tightened, install the crossbars in pairs, centering them between the clamps already in place. Insert carriage bolts through the crossbar slots, using washers and wing nuts to tighten the jig snug against the panel *(above)*. Then, tighten the bar clamps completely.

Wedged Clamping Bar

Building the jig body

The wedged clamping bar shown at left is an excellent alternative to a bar clamp for edge gluing boards, because it prevents the stock from bowing when pressure is applied. Cut the top and bottom from ¾-inch-thick stock, making them longer than the widest panel you will glue up. Cut the spacer, tail block, and wedges from stock the same thickness as the boards to be glued. (Keep sets of spacers, tail blocks, and wedges on hand to accommodate boards of varying thickness.) Use a machine bolt, washer, and wing nut at each end of the jig to secure the top, bottom, and spacers together. Wax the bars to prevent adhesive from bonding to them.

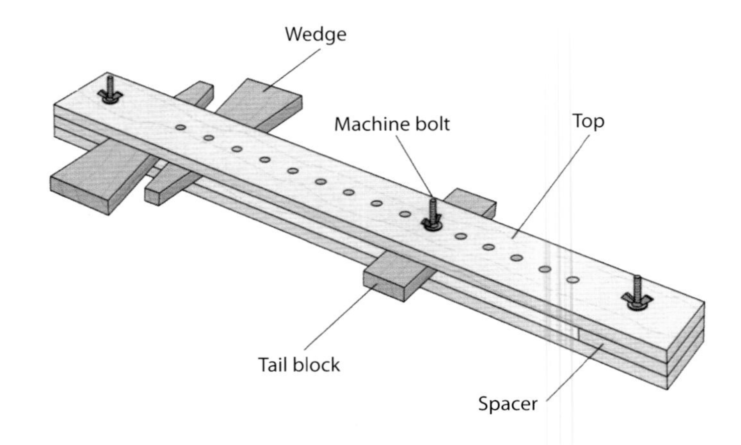

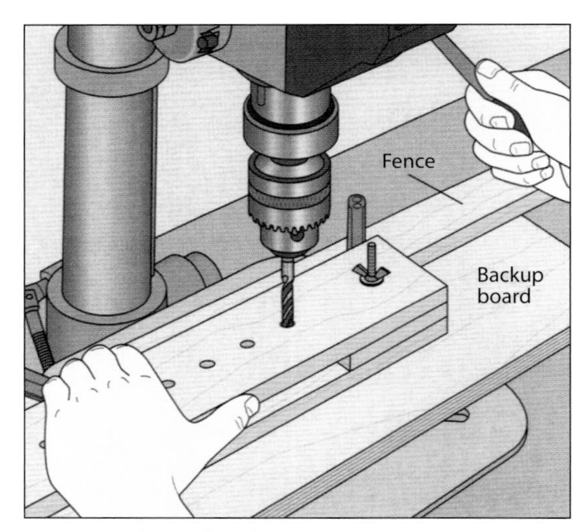

Preparing the jig for glueup

You need to bore holes through the jig to adjust it for the width of the panel to be glued. Since you will be drilling straight through the jig, clamp a backup board to your drill press table with a fence along the back edge to ensure the holes are aligned. Install a bit the same diameter as the machine bolt and place the tail block in place. Butt the jig against the fence and drill a hole through the top, the bottom, and the tail block. Bore the remaining holes through the jig body at 1½-inch intervals *(above)*.

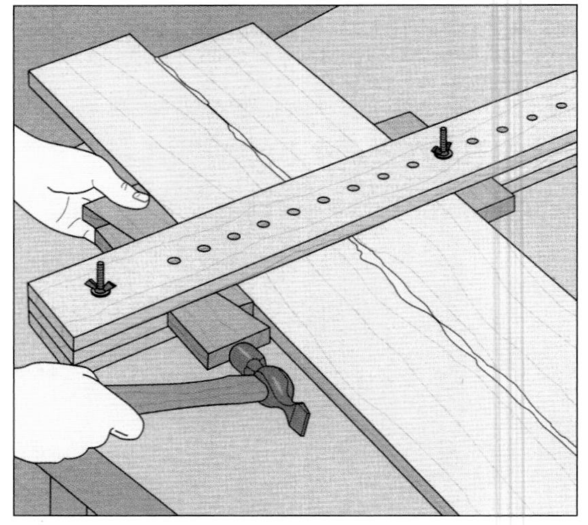

Edge gluing boards

Spread adhesive on the edges of your stock and set the boards face-down on a work surface. Slip a clamping bar over the boards and position it 6 to 12 inches from one end of the assembly. Butt the tail block against the far edge of the boards, using the machine bolt, washer, and wing nut to fix it in place. To apply clamping pressure, tap one of the wedges at the front edge of the panel *(above)* until there are no gaps between the boards and a thin glue bead squeezes out of the joints. Install the bars at 18- to 24-inch intervals.

A Jig for Clamping Thin Stock

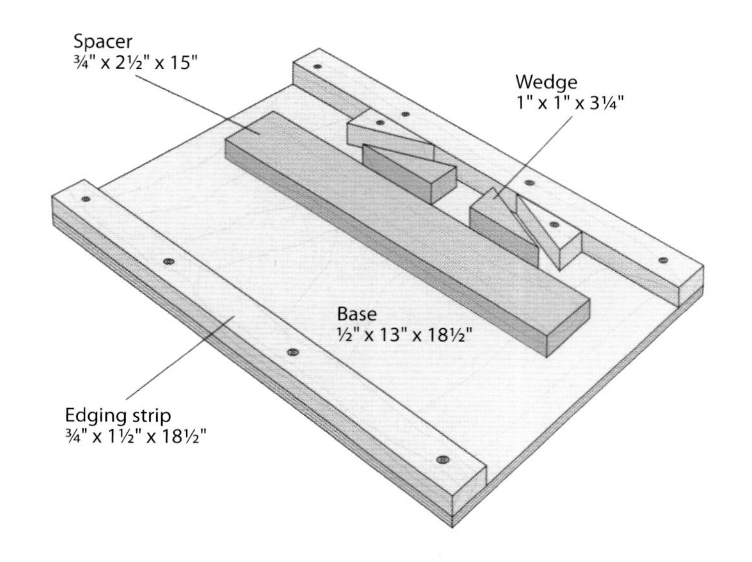

Edge gluing thin stock

The benchtop jig shown at right allows you to apply the correct clamping pressure for edge gluing thin stock. Cut the base from ½-inch plywood and the remaining pieces from solid stock. Refer to the illustration for suggested dimensions, but be sure the base is longer than the boards to be glued and the spacer is long enough to butt against the entire front edge of the panel. The edging strips should be thicker than your panel stock. Screw them along the edges of the base and fasten two wedges flush against one strip with their angled edges facing as shown at right. Wax the top face of the base to keep the panel from adhering to it. Apply glue to your stock and set the pieces on the base, butting the first board against the edging strip opposite the wedges. Butt the spacer against the last board and slide the two loose wedges between the spacer and the fixed wedges. Tap the wedges tight to apply clamping pressure *(below)*.

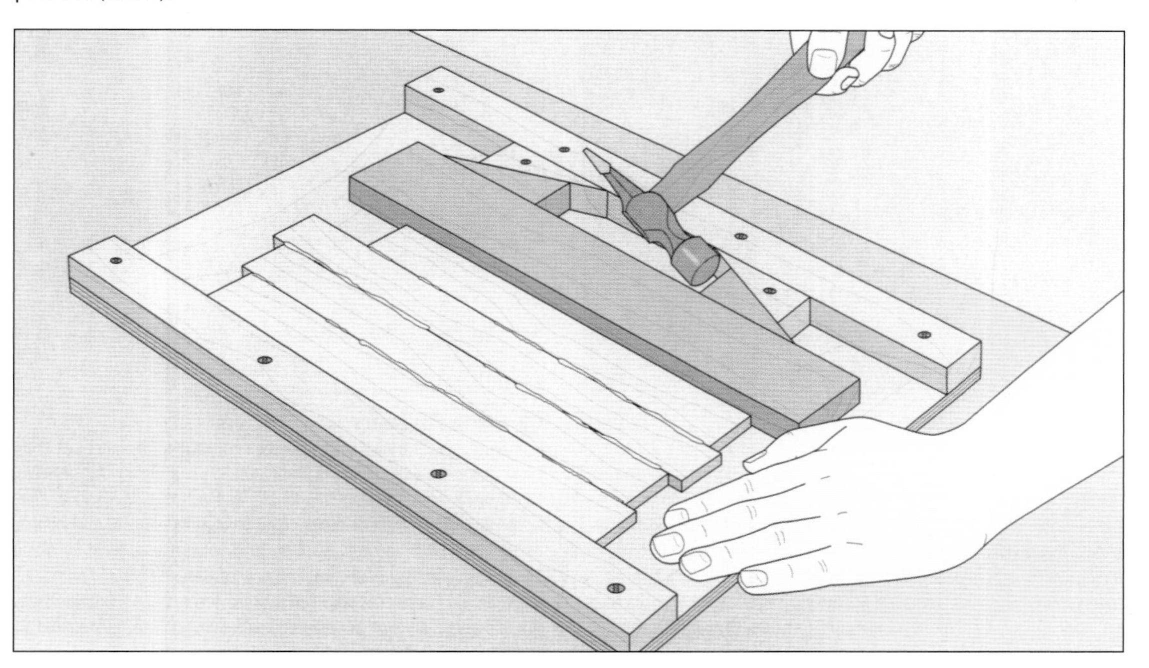

A Wall-Mounted Glue Rack

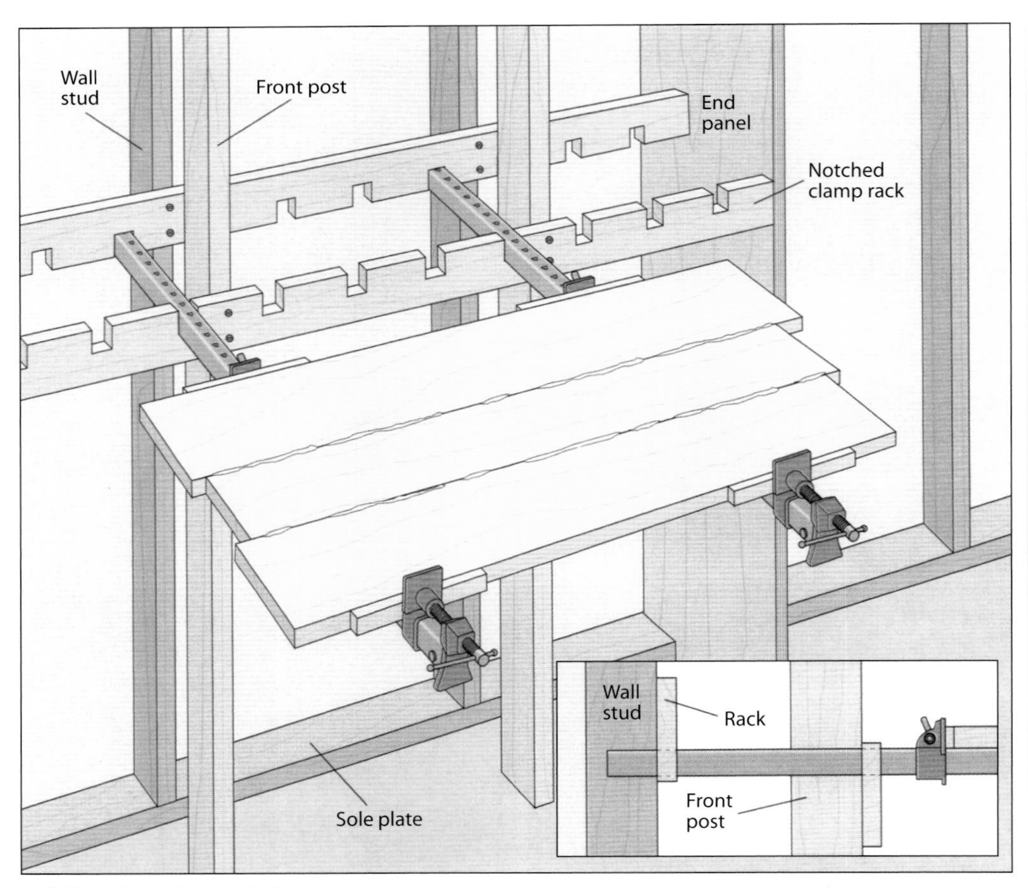

Building the rack and gluing up a panel

The jig shown above allows you to glue up panels using bar clamps, but saves shop space by being mounted to a wall. For clarity, the illustration shows only one pair of clamp racks, but you can install as many as you like from floor to ceiling at 12-inch intervals. Cut the clamp racks from 8-foot-long 1-by-4s and saw notches along one edge of each piece as you would for a sawhorse rack *(page 106)*. Attach one rack of each pair to the wall, driving two screws into every wall stud; make sure the notches are pointing down. To support the front clamp rack, cut floor-to-ceiling 2-by-4s as posts and position one directly facing each stud about 8 to 10 inches from the wall. Screw the front rack to these posts, positioning the notches face-up so they will hold the clamps level. Next, mount two ¾-inch plywood end panels to fit around the jig. Notch the bottom end of the panels to fit over the sole plate and fasten the top to the ceiling. Drive screws through the sides of the end panels into the ends of the racks. To use the jig to glue up a panel, slide bar clamps through the notches in the front and back racks, making sure the ends of the clamps extend beyond the stud-mounted rack *(inset)*. The rest of the operation is identical to edge gluing with any other clamp rack.

Frame-Clamping Jigs

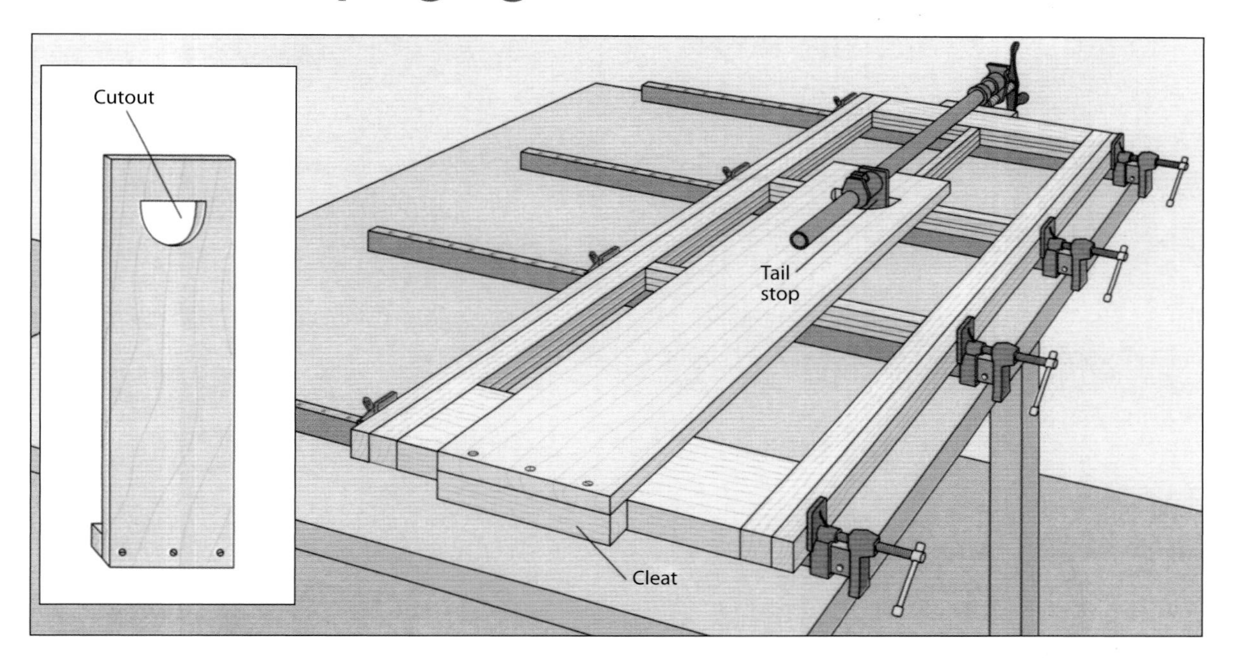

A Pipe Clamp Extender

Gluing up a large frame
The jig shown above will extend the capacity of your pipe clamps. Cut the main body of the extender from 1-by-6 stock and the cleat from a 2-by-2. Saw a D-shaped cutout near one end of the body to accommodate the pipe clamp tail stop, then screw the cleat to the opposite end *(inset)*. To apply clamping pressure on the top and bottom rails of a long frame like the one in the illustration, set the cleat against one end of the workpiece and fit the pipe clamp tail stop into the cutout. Then tighten the clamp so that the handle-end jaw is pressing against the opposite end of the workpiece *(above)*. Use wood pads to protect the workpieces.

Shop Tip

Doubling up pipe clamps
Another way to extend the capacity of shorter pipe clamps is to use them in pairs to function as a single long one. Set up the workpiece (here four boards to be edge glued) as you would on a glue rack *(page 106)*. To fashion a long clamp, position two shorter clamps across the workpiece so that the handle-end jaws rest against opposite edges and the tail stops of the clamps overlap. As you tighten one of the clamps, it will pull the joints together.

Work Supports

Supporting long planks and large panels as they are fed across a saw table is very cumbersome. Outfeed tables can be attached to most saws, but they tend to take up a lot of floor space.

Commercial roller stands, like the one shown in the photo at right, make better use of shop space; they can also be moved easily to where they are needed and adjusted to whatever height is suitable. The shop-made stands described below and on the following page share the advantages of the store-bought variety, with the additional benefit of being easy and inexpensive to build. They can also be dismantled and stored when not needed.

There are other work-support jigs that make life easier in the shop. The vise extension stand shown on page 114, for example, solves the problem of keeping long boards edge-up in a bench vise.

A commercial roller stand supports a board being ripped on a radial arm saw. The stand should typically be set ¼-inch below the level of the saw table and positioned two feet from its edge.

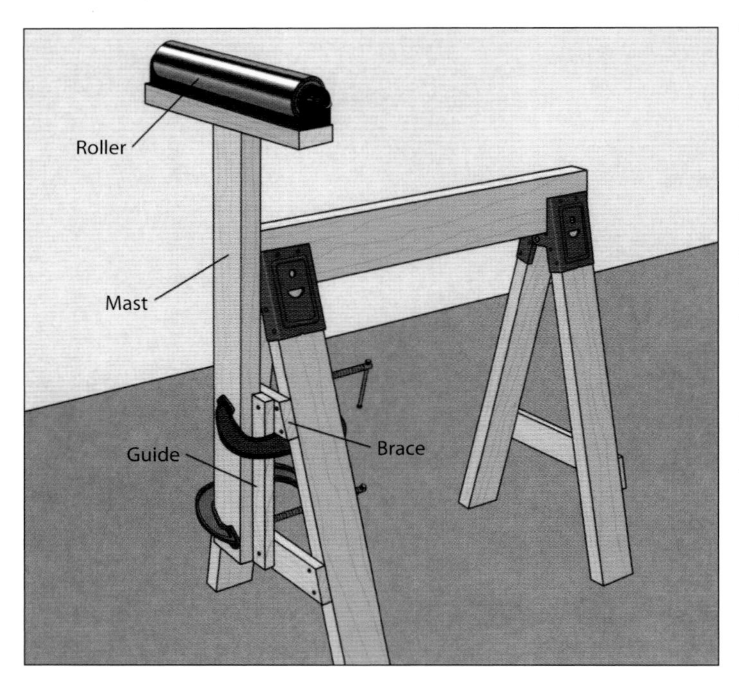

Two Shop-Made Roller Stands

Setting up a temporary stand

With only a sawhorse, two C clamps, and a commercial roller, you can make a simple roller stand like the one shown at left. Make a T-shaped mast for the roller that is long enough to hold it at a suitable height. Screw the roller to the horizontal part of the mast. Add a brace to the side of the horse for clamping the mast in place: Cut a 1-by-4 to span the legs between the sawhorse bracket and the original brace and screw it to the legs. Cut a 1-by-2 to span the two braces and screw it in place as a vertical guide for the mast. To secure the roller stand to the sawhorse, clamp the mast to the braces, making sure it is flush against the guide.

Building an adjustable roller stand
To build the roller stand shown at right, start by constructing the frame for the roller, cutting the four pieces from 1-by-4 stock. Glue the frame together with butt joints, adding screws to reinforce the connections. Then bore a hole in the middle of each side of the frame for a ¼-inch-diameter carriage bolt. Locate the hole 3 inches from the bottom of the frame. Insert the bolts from the inside of the frame and screw the roller to the top. As well as the commercial roller shown, two variations that permit you to feed the workpiece from any direction are shown below. Cut the remaining pieces of the stand from 1-by-6 stock, referring to the dimensions provided, then rout a ¼-inch-wide slot down the middle of the two uprights; the slot should be about 14 inches long. Screw the crosspiece to the uprights, aligning the top of the piece with the bottom of the slot. Fasten the uprights and rails to the feet. To guide the roller frame, nail 1-by-1 cleats to the uprights about ¼ inch in from the edges. To set up the stand, position the roller frame between the uprights, fitting the carriage bolts into the slots. Slip washers on the bolts and tighten the wing nuts to set the height of the roller.

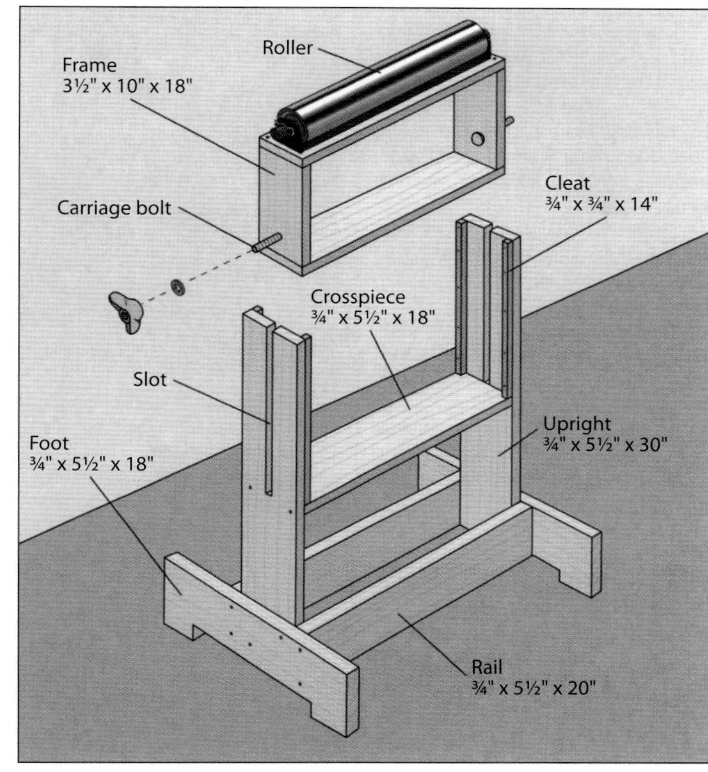

Types of Rollers

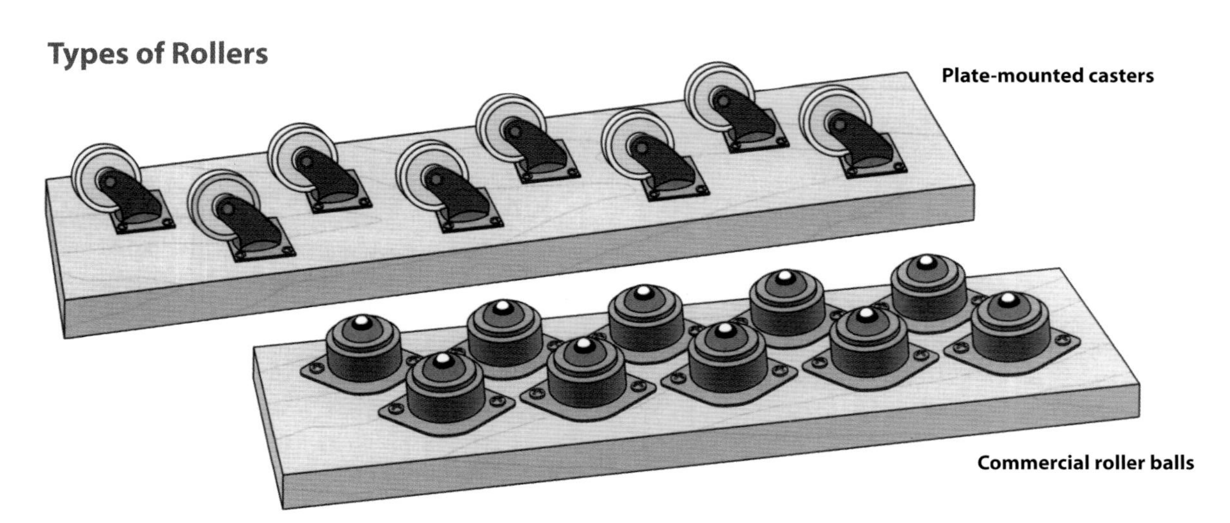

A Vise Extension Stand

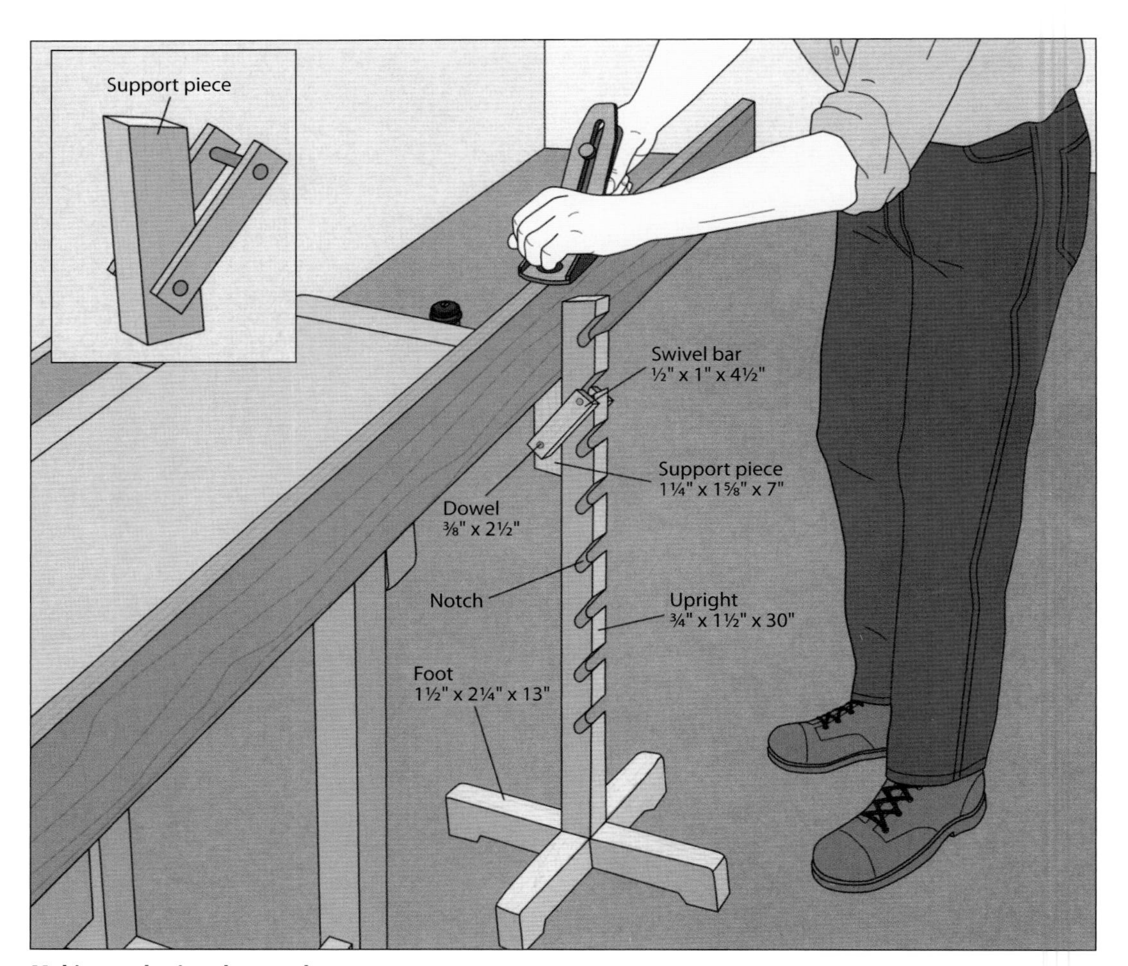

Making and using the stand

Also known as a bench slave, a vise extension stand is used to support the free end of a long board clamped in the shoulder vise of a workbench. Refer to the dimensions in the illustration for a stand that works well with most workbenches.

To build the stand, cut the upright to length and, starting 5 inches from the bottom, saw angled notches at 2½-inch intervals along its length. Cut the notches about 1 inch long and ½ inch wide. Then saw the feet to length and cut recesses along their bottom edges. Join the feet with a cross lap joint: Cut a lap in the top edge of one foot and in the bottom edge of the other foot. Glue the two feet together. Once the adhesive is dry, screw the upright to the feet. Cut the support piece and swivel bars, angling the top of the support piece about 10° *(inset)*. To join the support piece to the swivel bars, bore holes for ⅜-inch-diameter dowels through the piece and near the ends of the bars, and slip the dowels into the holes; glue them in place. To use the stand, insert the dowel at the top end of the swivel bars in the appropriate slot in the upright for the height you need and prop your workpiece on the support piece.

Cutting Large Panels on the Band Saw

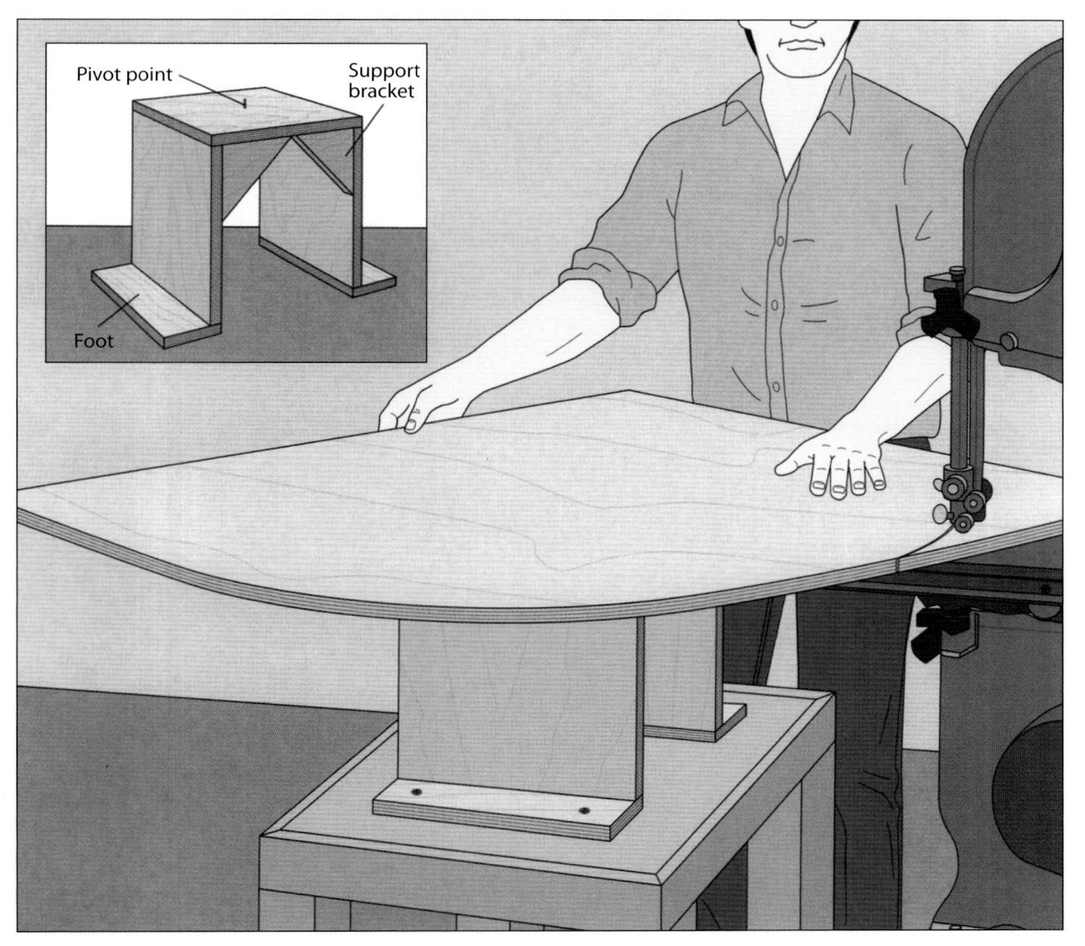

Making and using the jig

For making circular cuts out of large panels on the band saw, use a jig like the one shown above. Build the jig from ¾-inch plywood, cutting the pieces so the top of the jig is level with the saw table when the feet are screwed or clamped to a work table. Before assembling the jig, drive a 1¼-inch-long screw as a pivot point through the center of the top piece so the tip of the screw projects from the surface by about ½ inch *(inset)*. Then screw the top and feet to the sides of the jig, and attach the triangular-shaped support brackets to the top and sides; be sure to countersink the fasteners.

Before setting up the jig, mark the center and circumference of the circle on the workpiece. Then cut from the edge of the piece to the marked circumference and back to the edge, creating a starting point for the circular cut. Now set up the jig: Attach it to a table and place the workpiece on the jig so the marked center of the circle contacts the pivot point. Position the table so the blade butts against the marked circle and the pivot point is aligned with the center of the blade and the machine's center line. Cut the circle by rotating the workpiece into the blade.

Tool Stands and Tables

A stand or table can transform a portable power tool into a reasonable facsimile of a full-size stationary machine. What they concede in power to their larger cousins, bench-mounted tools compensate with portability, ease of storage, and lower price.

There are commercial stands for benchtop tools, but you can easily build a stand like the one shown below. Storage shelves and drawers can be added to customize the basic design. There is one requirement, however: Ensure the stand's surface area is large enough for your needs and that it supports the tool at a comfortable height. The extension router table shown opposite not only converts a router into a mini-shaper, but can be easily removed when it is not needed. A more elaborate, but versatile option is illustrated on page 118. The three-in-one portable power tool table features replaceable inserts for a router, an electric drill, and a saber saw.

Because of its central role in woodworking, the router merits a dedicated table in most shops. The shop-built benchtop version illustrated on page 120 allows you to take advantage of this tool's great versatility.

Supports for Portable Power Tools

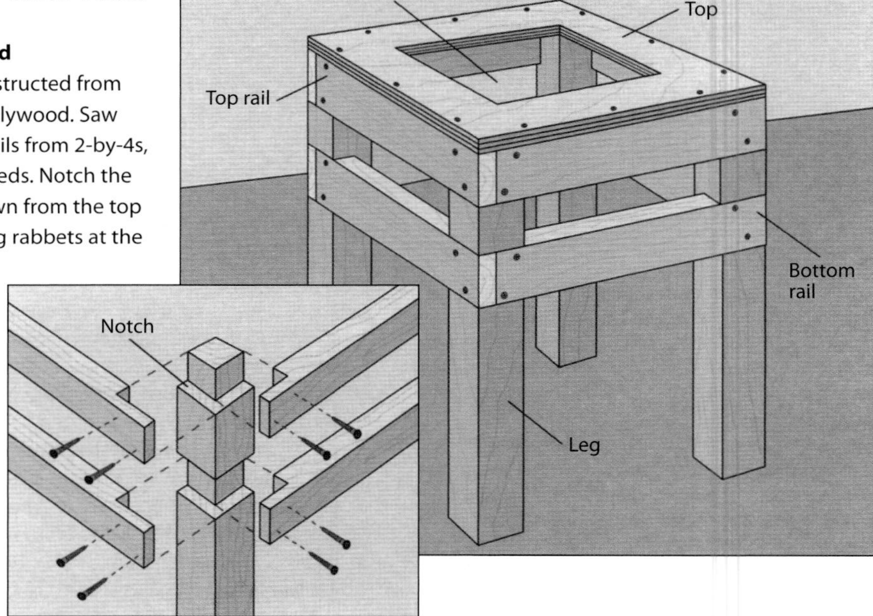

Building a benchtop tool stand

The stand shown at right is constructed from 4-by-4 and 2-by-4 lumber and plywood. Saw the legs from 4-by-4s and the rails from 2-by-4s, sizing the pieces to suit your needs. Notch the legs at the top and 6 inches down from the top to fit the rails, then cut matching rabbets at the ends of all the rails *(inset)*. Glue up the legs and rails, adding countersunk screws to reinforce the joints. Cut the top from ¾-inch plywood. If you plan to place a table saw on the stand, saw a square hole out of the center of the top as shown to allow sawdust to fall through; place a box underneath to catch the waste. Finally, screw the top to the legs and rails, again countersinking the fasteners. When using a tool on the stand, secure it to the top with screws or clamps.

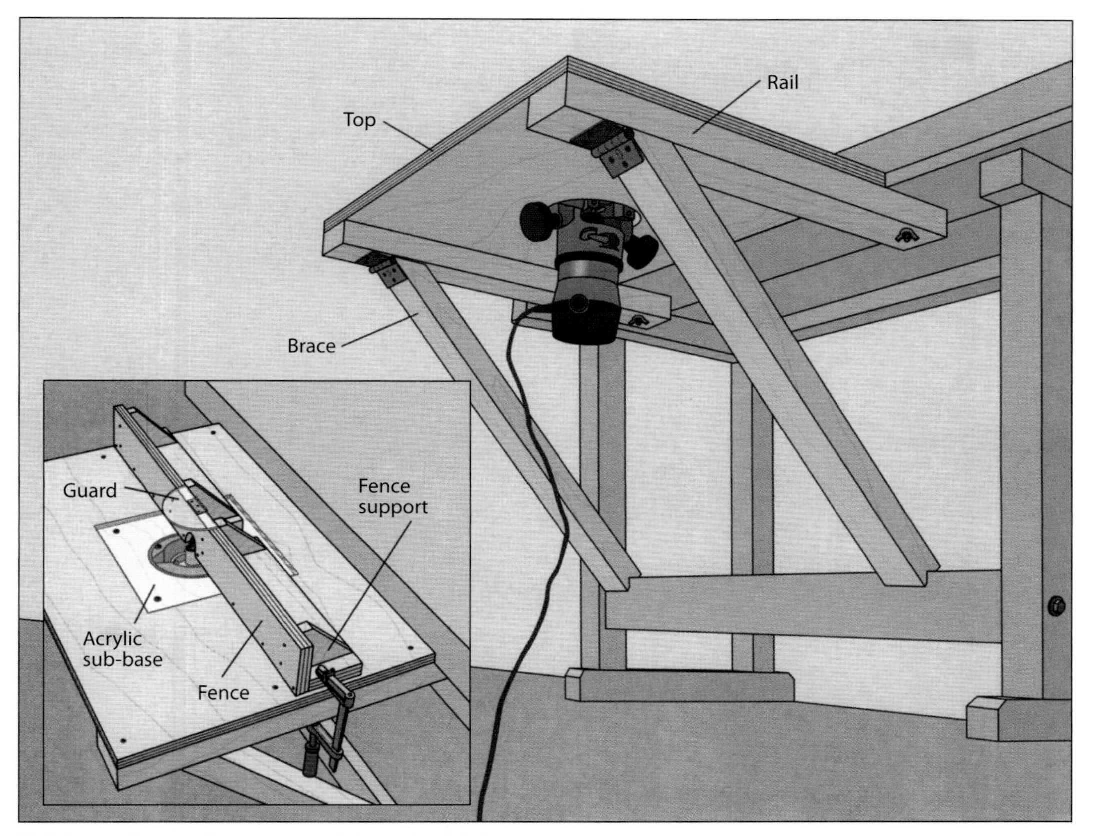

Making and mounting a removable router table

Attached to a workbench or table, the extension table shown above serves as a compact router table that can be stored when it is not needed. Size the parts according to your needs. Start by cutting the top from ¾-inch plywood, and the rails and braces from 2-by-4 stock. Saw the rails 6 inches longer than the width of the top so they extend under the top and can be fastened to the underside of the bench using nuts and hanger bolts. The hinged braces should be long enough to reach from the underside of the rails to a leg rail on the bench. Cut a bevel at the top end of the braces and an angled notch at the bottom end. The router is attached to the top with a square sub-base made of ¼-inch clear acrylic. Several steps are necessary to fit the base to the tabletop and then to the router. First, lay the square sub-base in the center of the table, clamp it in place, and mark its edges with a pencil. Mark the center of the sub-base and drill a pilot hole completely through the base and the tabletop. Remove the sub-base and turn your attention to the tabletop. Use your router to plow a ¼-inch-deep recess within the pencil outline of the sub-base. Then, using the pilot hole as a center and your router as a template, cut a round hole through the tabletop the size of your router's standard base. The tabletop is now ready. In the sub-base, drill a hole in the center that is slightly larger than your largest router bit, and screw the base to the router, using countersunk machine screws. Lay the sub-base in the table recess and screw it down, countersinking the wood screws. All surfaces should be flush. For a fence, cut two pieces of ¾-inch plywood and screw them together in an L shape. Saw a notch out of the fence's bottom edge to accommodate your largest bit, then screw on four fence supports for added stability. Attach a clear semicircular plastic guard with a hinge to allow it to be raised out of the way *(inset)*. The fence is clamped in place.

Portable Power Tool Table

Easy and inexpensive to build, the versatile table shown below allows you to convert three different portable power tools into stationary tools: the electric drill, the router, and the saber saw. The table features a spacious table-top, an adjustable fence, a storage shelf, and a conveniently located on/off switch. The tabletop includes a rectangular cutout to accept a custom-made insert for each of the three power tools.

Use ¾-inch plywood for the table-top, the shelf, the cleats, the inserts and the fence; ¼-inch plywood for the support brackets; and solid lumber for the other parts (2-by-4s for the legs and 1-by-3s for the rails). Refer to the cutting list for suggested dimensions.

Start building the table by preparing the tabletop for the tool inserts. Cut a rectangular hole out of its center the same size as the inserts. Then screw the cleats to the underside of the

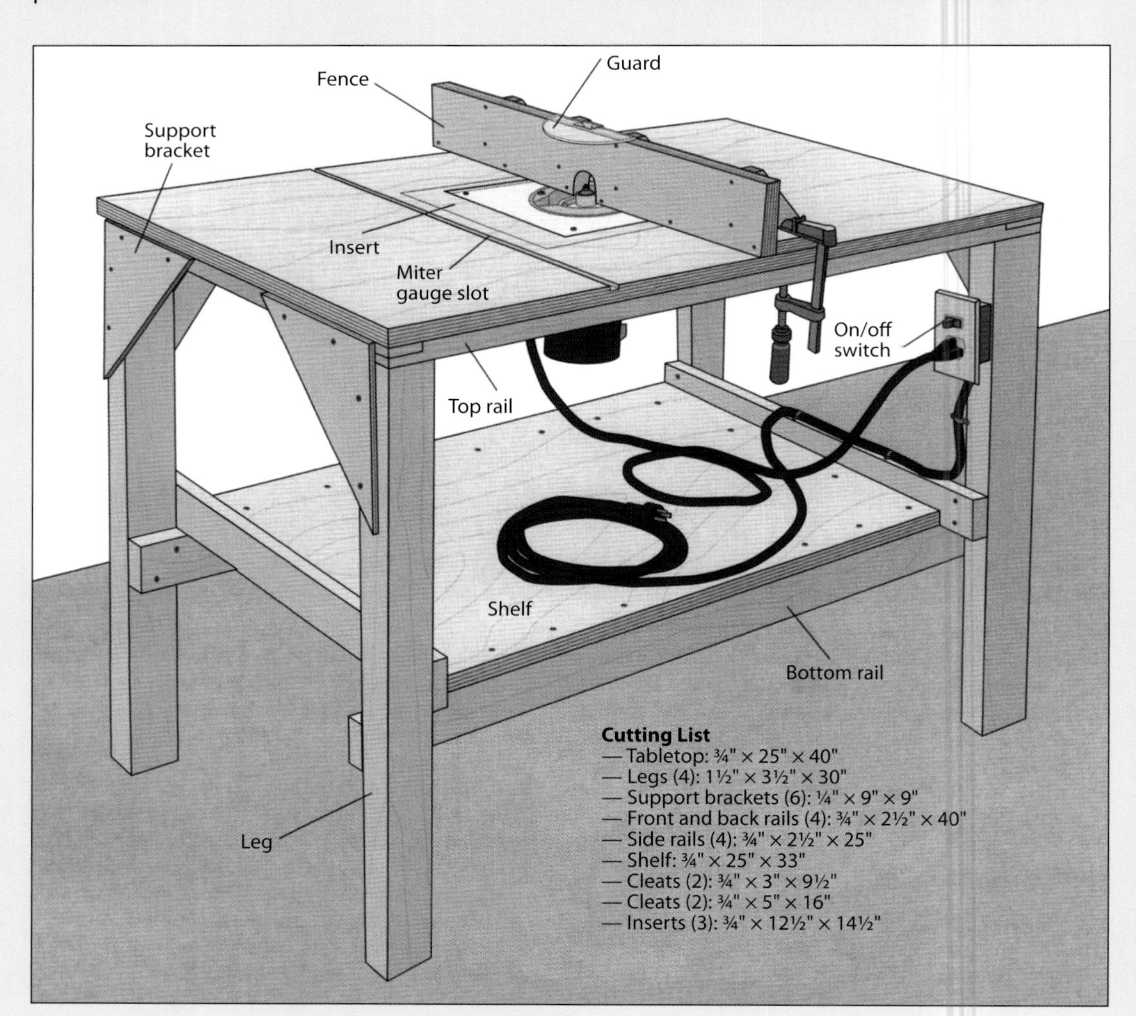

Cutting List
— Tabletop: ¾" × 25" × 40"
— Legs (4): 1½" × 3½" × 30"
— Support brackets (6): ¼" × 9" × 9"
— Front and back rails (4): ¾" × 2½" × 40"
— Side rails (4): ¾" × 2½" × 25"
— Shelf: ¾" × 25" × 33"
— Cleats (2): ¾" × 3" × 9½"
— Cleats (2): ¾" × 5" × 16"
— Inserts (3): ¾" × 12½" × 14½"

top, forming a ledge to which the inserts can be fastened *(below)*.

Before assembling the table, rout a ⅜-inch-deep dado across the table about 12 inches from the left-hand end to accommodate a miter gauge. Then screw the parts of the table together. Use lap joints for the top rails (placed flat), then screw this frame onto the top of the legs. Screw the bottom rails (placed on edge) to the legs, then attach the shelf. You can either countersink the fasteners or counterbore the holes, and then conceal the screw heads with wood plugs.

Next, saw the three tool inserts, sizing them to fit precisely in the hole in the tabletop. Prepare the router insert as you would the top of the removable router table shown on page 117. To mount the insert in the table, set it in place on the cleats and bore a hole through the insert and the cleats at each corner; the holes should be countersunk. Screw the insert to the cleats.

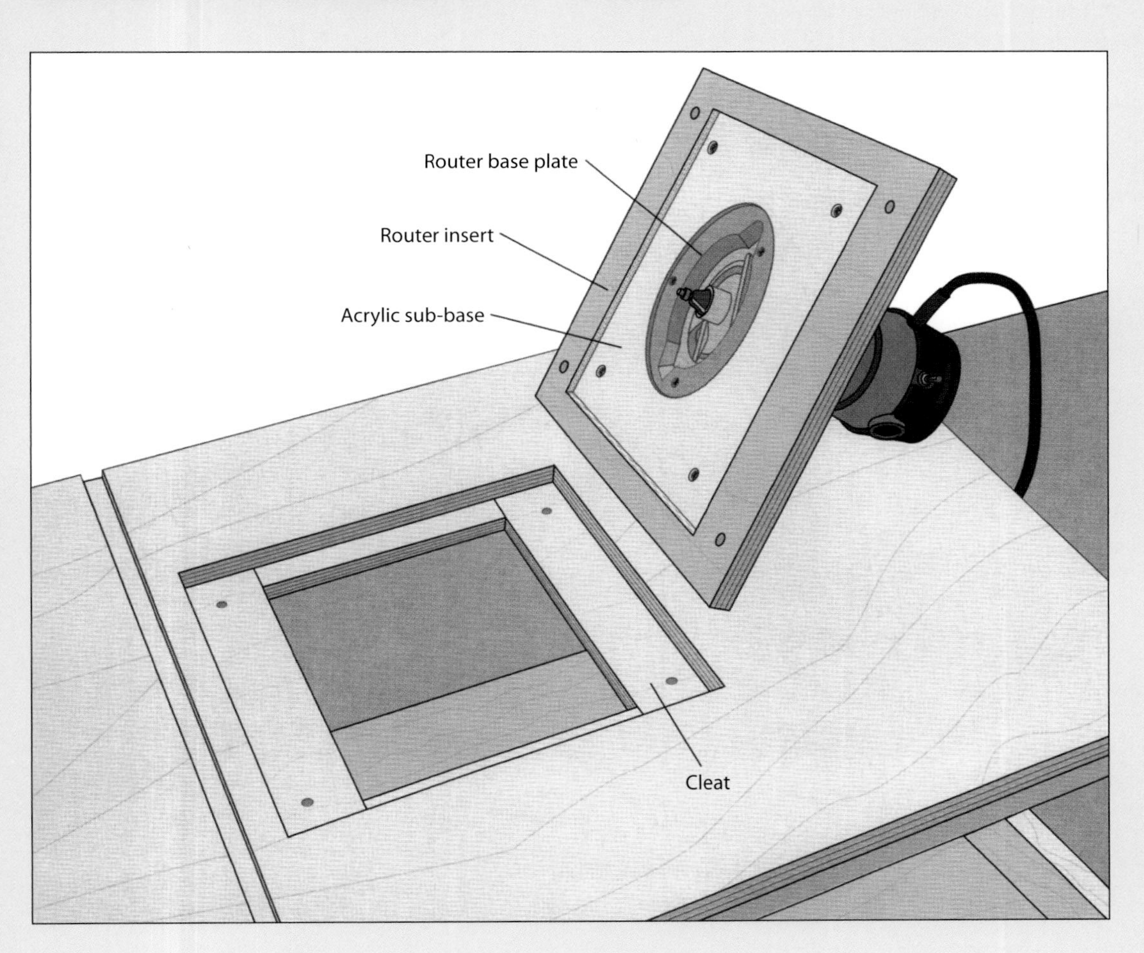

Tables for Power Tools

A Router Table/Cabinet

Assembling the table
Built entirely from ¾-inch plywood, the table shown at left allows you to use your router as a stationary tool—a requirement of many operations. It features a large top with a slot for a miter gauge, an adjustable fence, a storage shelf, and cupboards. Start with the basic structure of the table, sizing the bottom, sides, back, shelf, dividers, and doors to suit your needs. Fix these parts together, using the joinery method of your choice. The table shown is assembled with biscuit joints and screws. Bore a hole through the back panel for the switch's power cord. For the top, cut two pieces of plywood and use glue and screws to fasten them together; the pieces should be large enough to overhang the sides of the cabinet by 2 or 3 inches. Fix the top to the cabinet. Finally, fasten a combination switch-receptacle to one of the dividers, with a power cord long enough to reach an outlet.

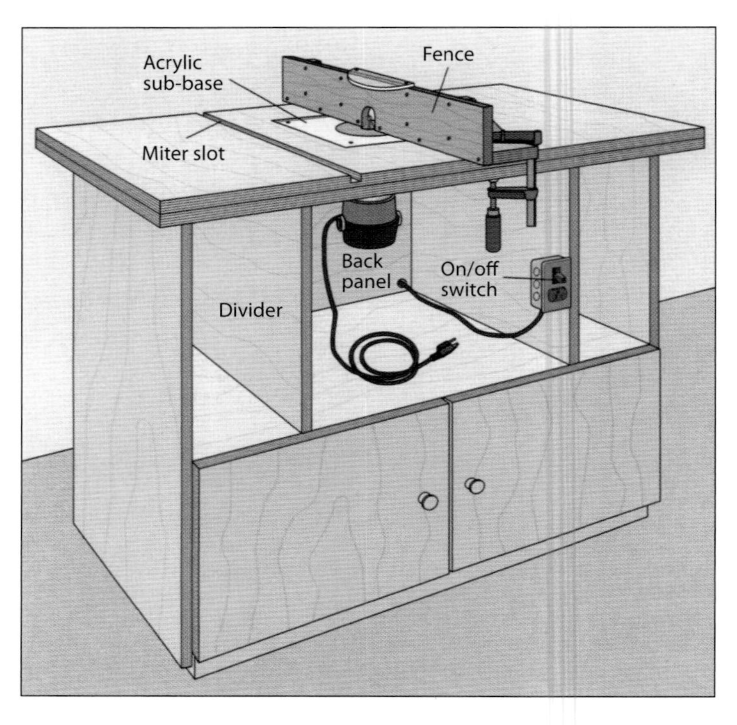

Recess

Top

Preparing the tabletop
The router is attached to the top with a square sub-base of ¼-inch-thick clear acrylic. Several steps are necessary to fit the sub-base to the top and then to the router. First, position the sub-base at the center of the top and outline its edges with a pencil. Mark the center of the sub-base and drill a pilot hole through the acrylic and the top. Remove the sub-base and rout out a ¼-inch-deep recess within the outline *(left)*. Use a chisel to pare to the line and square the corners. Then, using the pilot hole as a center, cut a hole through the top to accommodate your router's base plate. Next, use a straightedge guide to help you rout the miter slot across the top. Clamp the guide square to the front edge of the top and butt the router against it as you plow a slot that is just wide enough to fit your miter gauge bar snugly.

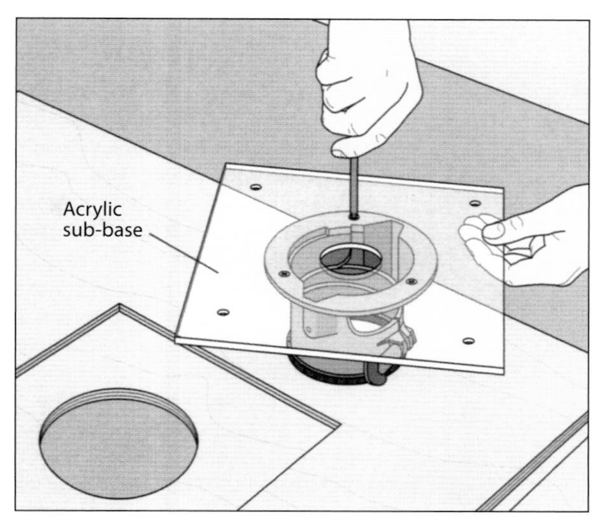

Preparing the sub-base

Drill a hole through the center of the sub-base slightly larger than your largest router bit, and fasten the sub-base to the router using flat-head machine screws *(above)*. Set the sub-base in the table recess and attach it with wood screws, drilling pilot holes first and countersinking all fasteners.

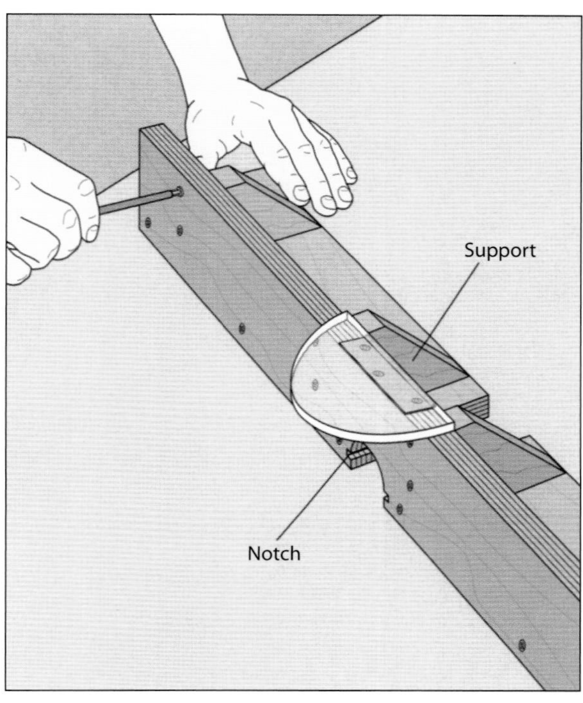

Shop Tip

A router table on the table saw

To make the most of the space in your shop, build a router table into your table saw's extension table. Rout a ¼-inch-deep recess into a non-metallic section of the top and cut a piece of acrylic to fit into the depression. With a saber saw, cut a hole in the recess to accommodate your router's base plate. Then remove the sub-base from the tool, screwing the router to the plastic piece instead. Next screw the plastic into the recess; countersink all the fasteners. Reattach the router to the base plate. A fence can be cut from plywood and attached to the saw fence when necessary.

Making the fence

Cut two pieces of ¾-inch plywood and screw them together in an L shape. Saw a notch out of the fence's bottom edge to accommodate your largest bit, and screw four triangular supports to the back for added stability *(above)*. Attach a clear semicircular plastic guard, with a hinge so it can be swung out of the way. To use the router table for a straight cut, clamp the fence in position and feed the workpiece into the bit, holding it flush against the fence.

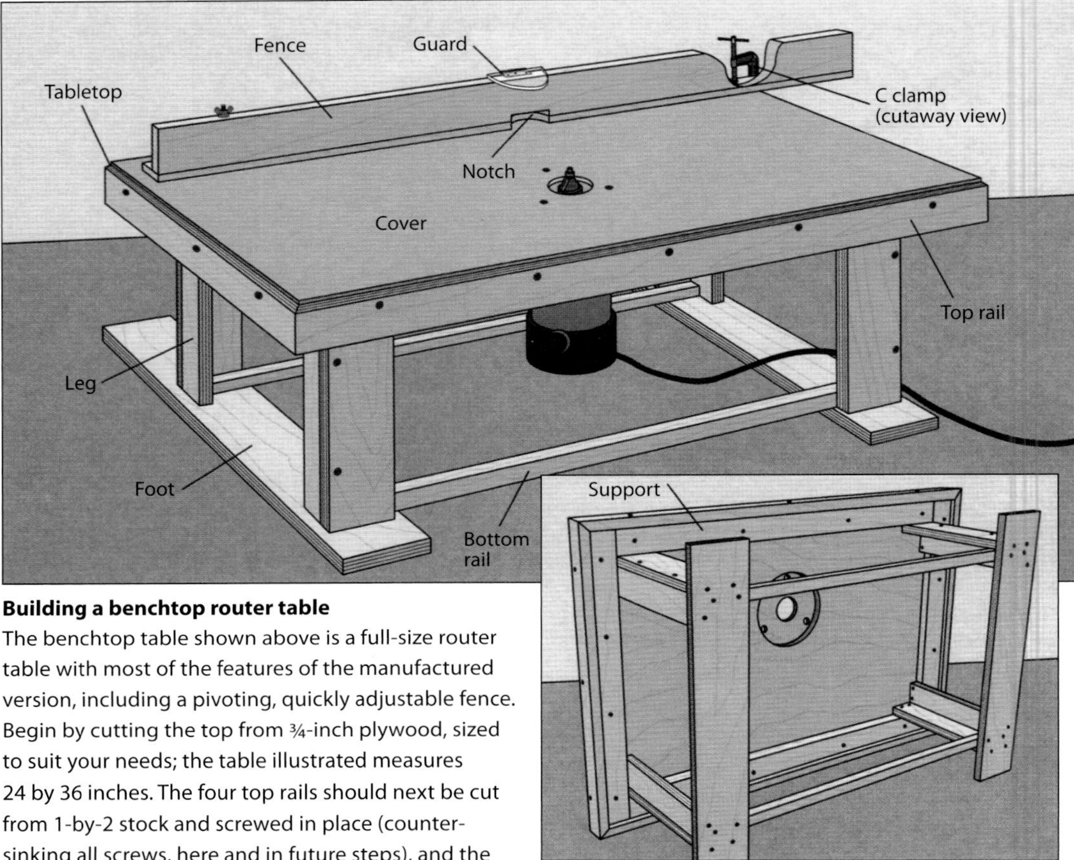

Building a benchtop router table

The benchtop table shown above is a full-size router table with most of the features of the manufactured version, including a pivoting, quickly adjustable fence. Begin by cutting the top from ¾-inch plywood, sized to suit your needs; the table illustrated measures 24 by 36 inches. The four top rails should next be cut from 1-by-2 stock and screwed in place (countersinking all screws, here and in future steps), and the entire top should be covered with a piece of ¼-inch plastic laminate, chamfered at the edges. Turn the table over so you can screw supports around the inside edges and attach the legs to the rails and top. The supports, legs, and feet can be constructed of ¾-inch plywood; the final dimensions will be determined by the size of your table. Make sure the legs are at least long enough to furnish ample room for your router. To prepare the tabletop for the router, drill a hole about 8 inches from the front center; make it slightly larger than your largest router bit. On the underside of the top, center the router over the hole and trace its outline. Use the router to plow a ¼-inch recess within the outline to accommodate your router base plate (or, make an acrylic sub-base and mount it as shown on page 121). Mark the location of the base plate screw holes, drill counterbore holes, and you will be ready to fasten the router in place. Next, construct the fence (about 6 inches longer than the top) out of two pieces of 1-by-3 stock screwed together in the form of an L. Through the base of the L, drill a hole for a ¼-inch carriage bolt about six inches from one end. Now center the fence about 6 inches from the rear of the top, mark the position of the hole, and drill for the carriage bolt. Slip a bolt through the hole; using that as a pivot, swing the right end of the fence forward. When the fence reaches the hole you cut for the router bit, mark the hole's position on the fence. That is where you will cut a clearance notch to accommodate your largest router bit. Make a guard that is hinged so it will swing out of the way like the one shown on page 117. Screw your router to the top and assemble the fence by inserting the carriage bolt from the bottom, using washers and wing nuts to tighten it. Adjust the fence for any width of work by pivoting it into place and securing the free end with a C clamp.

A Band Saw Extension Table

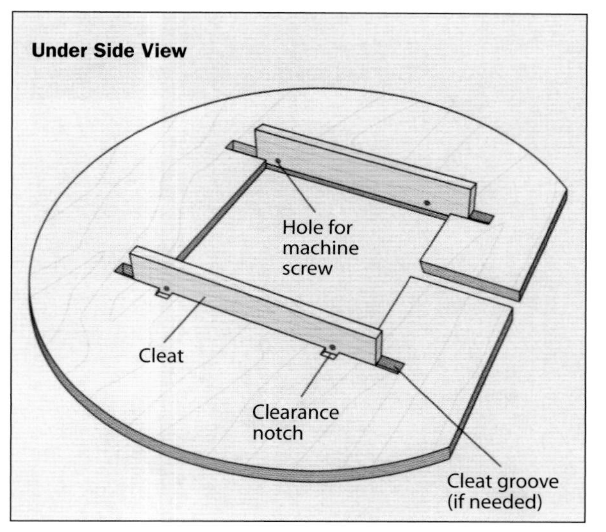

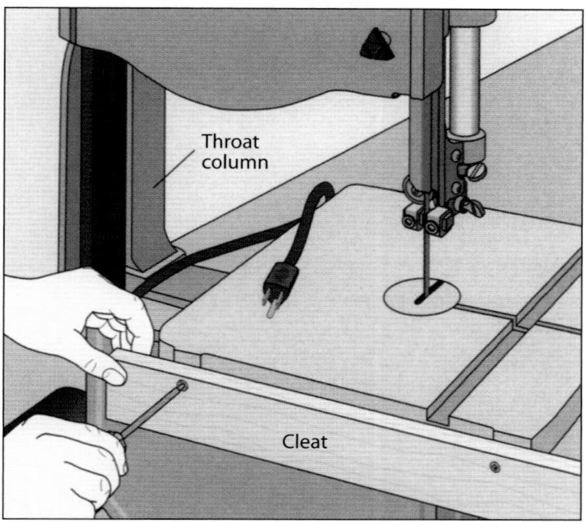

Building and installing a table for the band saw
An extension table on your band saw will enable you to cut long or wide pieces with greater ease and control. Using ¾-inch plywood, cut the jig top to a suitable diameter, then saw out the center to fit around the saw table and trim a portion of the back edge to clear the throat column. Cut a 1½-inch-wide channel from the back of the table to the cutout so the table can be installed without removing the blade. Next, prepare two cleats that will be used to attach the jig to the saw table. For these, two 1-by-3s should be cut a few inches longer than the saw table. Position them against the sides of the saw table so that they are ¾ inch below the table surface, with at least ¼ inch of stock above the threaded holes. Depending on the position of the holes on your saw table, you may have to position the top of the cleats closer than ¾ inch to the top of the saw table. In that case you will have to rout grooves for the cleats on the underside of the top to allow the tabletop to sit flush with the machine's table *(above)*. Mark the hole locations on the cleats, bore a hole at each spot, and fasten the cleats to the table with the machine screws provided for the rip fence (right, top). Then place the tabletop on the cleats and screw it in place *(right, bottom)*; be sure to countersink the screws. The top should sit level with the saw table.

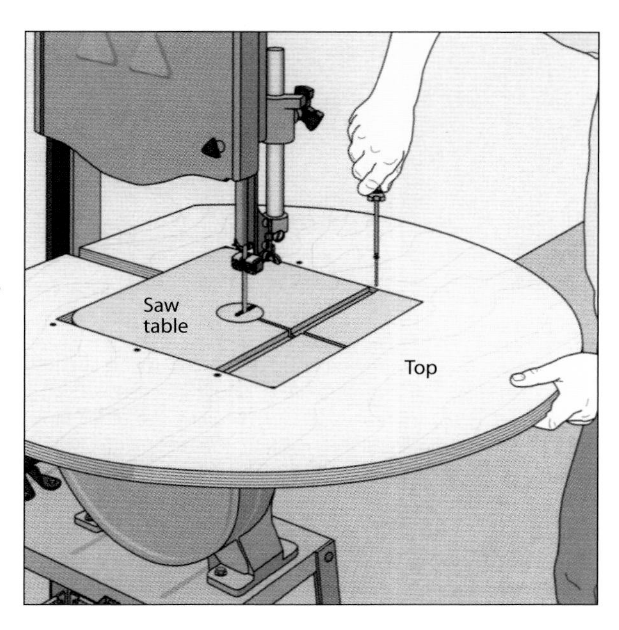

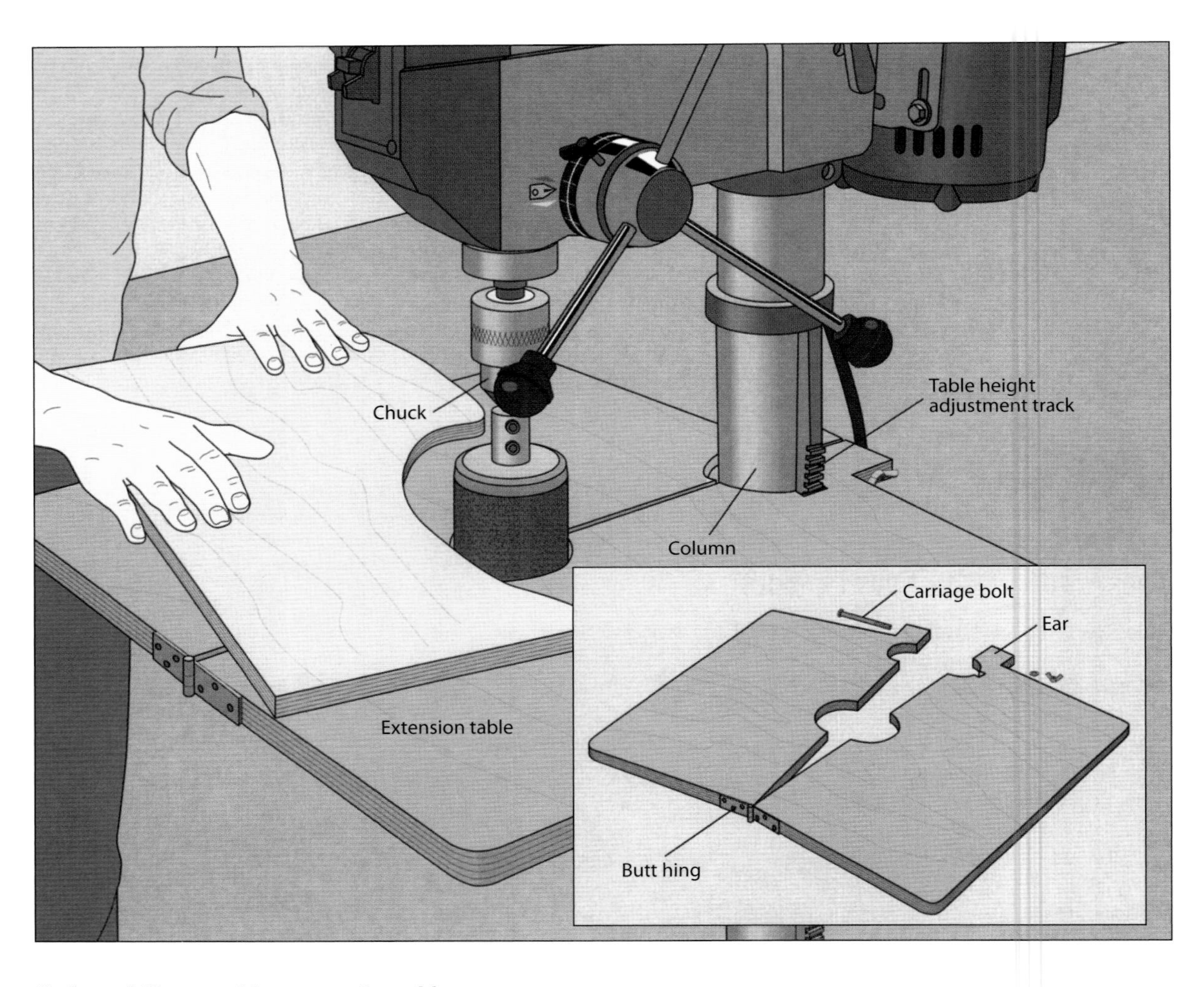

Fitting a drill press with an extension table

The small table typical of most drill presses will not adequately support large workpieces, especially when the tool is set up for sanding operations. To build an extension table, start by cutting a piece of ¾-inch plywood into a square with dimensions that suit your needs. Then mark a line down the middle of the piece and draw two circles centered on the line. Locate the first about 4 inches from the back edge, sizing it to fit snugly around the drill press column. Locate the second hole under the chuck; make its diameter about ½ inch greater than the largest accessory you plan to use. To help pinpoint the center of the hole, install a bit in the chuck and measure the distance from the column to the bit. Prepare to install the jig on the drill press table by cutting its back edge, leaving a rectangular "ear" that protrudes behind the back hole. Bore a hole through the ear for a ¼-inch-diameter carriage bolt. Next, saw the jig in half along the centerline and cut out the two circles. You may need to make other cuts to clear protrusions on your machine. On the model shown, a notch was needed for the table height adjustment track on the throat column. Finally, screw a butt hinge to the front edge of the jig to join the two halves as shown. The carriage bolt and wing nut will clamp the table in place on top of the drill press table.

Plate Joiner Stand

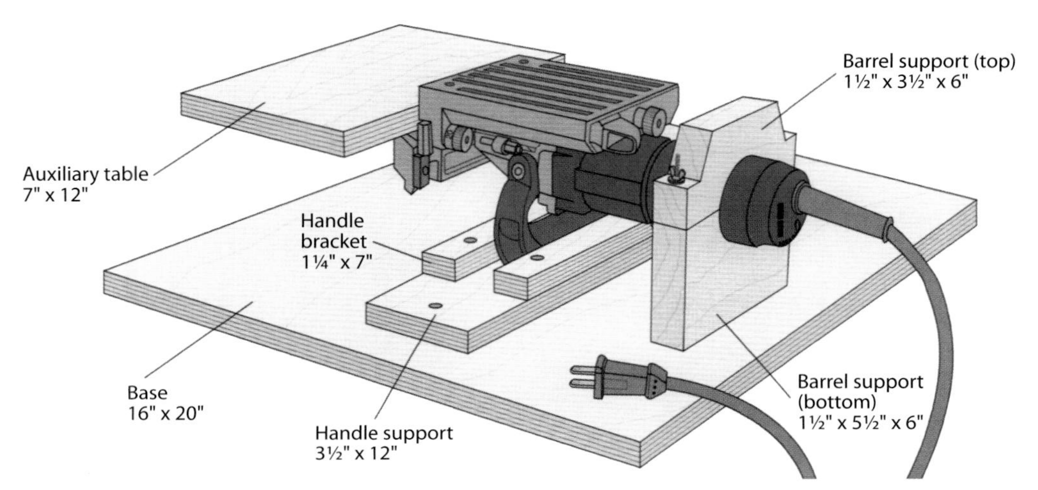

Building the jig

Paired with a plate joiner, the jig shown above will reduce the setup time needed to cut slots for biscuits in a series of workpieces. Build the jig from ¾-inch plywood, except for the barrel support, which should be solid wood. Refer to the illustration for suggested dimensions. Screw the handle support to the base and attach the handle brackets, spacing them to fit your tool. With the plate joiner resting upside-down on the handle support, butt the barrel support against the motor housing and outline its shape on the stock. Bore a hole for the barrel and cut the support in two across its width, through the center of the hole. Screw the bottom part to the base and fit the other half on top. Bore holes for hanger bolts through the top on each side of the opening, then drive the bolts into the bottom of the support. Use wing nuts to hold the two halves together. Finally, screw the auxiliary table to the joiner's fixed-angle fence. (It may be necessary to drill holes in the fence for the screws.)

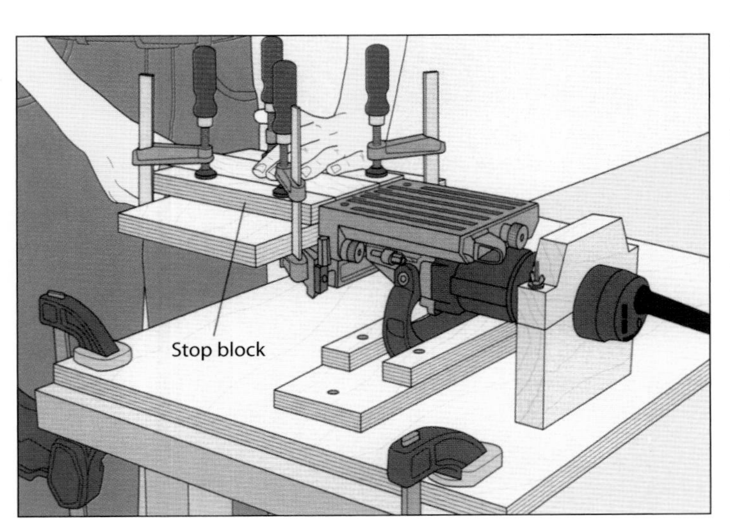

Cutting the slots

Secure the plate joiner in the stand and clamp the jig base to a work surface. Set the fence at the correct height and, for repeat cuts, clamp stop blocks to the auxiliary table to center the workpiece in front of the cutter. For each cut, put the workpiece flat on the table and butted against the joiner's faceplate, then turn on the tool and push the stock and the table into the blade *(left)*.

Extension Tables

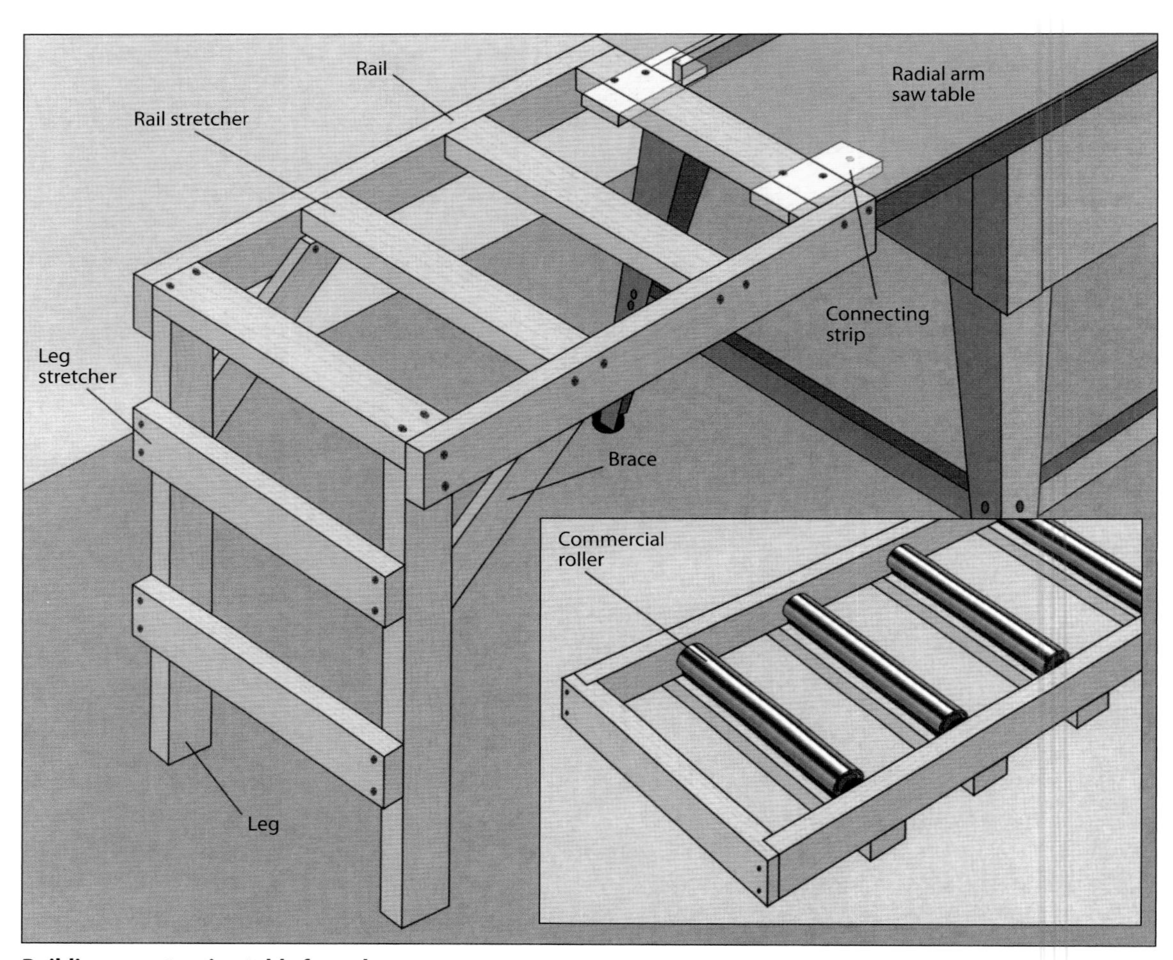

Building an extension table for a chop saw

Made entirely from 2-by-4 and 1-by-3 stock, the extension table shown above can be attached to the outfeed or infeed ends of a chop saw table. Using 2-by-4s, cut the legs, rails, and stretchers to suit the dimensions of your saw, making the length of the legs equal to the distance between the top of the saw table and the shop floor, less the thickness of the stretchers. Attach the rail stretchers so that their tops are flush with the rail's top edges. Attach the leg stretchers to the legs, then screw the legs to the inside edges of the rails. Make certain the outside rail stretcher is butted against the tops of the legs. Cut the braces from 1-by-3s to reach from the bottom of the second rail stretcher to the inside edges of the legs. Miter the ends of the braces and screw them in place. To fasten the extension table to the saw table shown, cut two wood strips and screw one end of each piece to the underside of the inside rail stretcher. Set the extension table flush against the saw table and fasten the other end of the strips to the underside of the table, using shims or spacers as needed. If you prefer to span the table's rails with rollers *(inset)*, rather than wood stretchers, cut the rail stretchers long enough to fasten them to the bottom edge of the rails. Then screw commercial rollers to the tops of the stretchers, placing shims under the rollers, if necessary, to set them level with the top of the saw table.

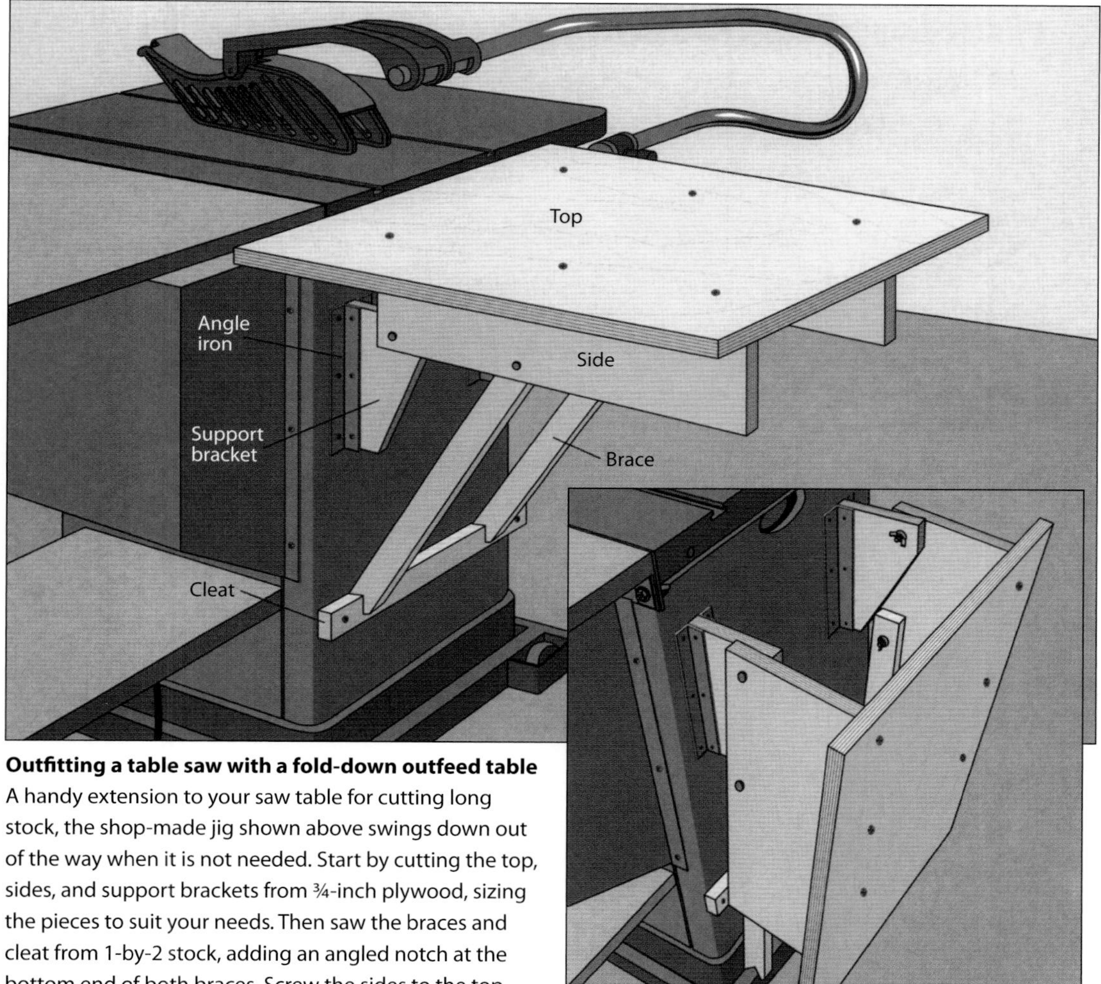

Outfitting a table saw with a fold-down outfeed table
A handy extension to your saw table for cutting long stock, the shop-made jig shown above swings down out of the way when it is not needed. Start by cutting the top, sides, and support brackets from ¾-inch plywood, sizing the pieces to suit your needs. Then saw the braces and cleat from 1-by-2 stock, adding an angled notch at the bottom end of both braces. Screw the sides to the top, countersinking the fasteners. Next, get ready to attach the jig to the saw housing. First, attach an angle iron to each side of both support brackets. Then, have a helper hold the top against the saw table, making sure the two surfaces are level; leave a slight gap between the top and saw table so the jig will fold down without jamming against the table. Now determine the position of the support brackets by butting each against the inside face of a side piece. Mark the holes in the angle irons on the saw housing. Drill a hole for a machine screw at each mark and fasten the angle irons to the housing. Reposition the jig against the saw table and bore holes for a carriage bolt through the sides and support brackets. Use washers under the nuts and bolt heads, and between the sides and brackets. Attach the braces to the sides with bolts spaced about 8 inches from the bracket bolts. Leave all the bolts loose enough for the sides and braces to pivot. Then, holding the jig level again, swing the braces toward the saw housing. Mark the points where the braces contact the housing and screw a cleat to the housing so the cleat's top surface aligns with the two points. To set the jig in position, rest the braces on the cleat. To fold the table down *(inset)*, raise the top slightly, move the braces off the cleat and swing the jig down.

A Spraying Turntable

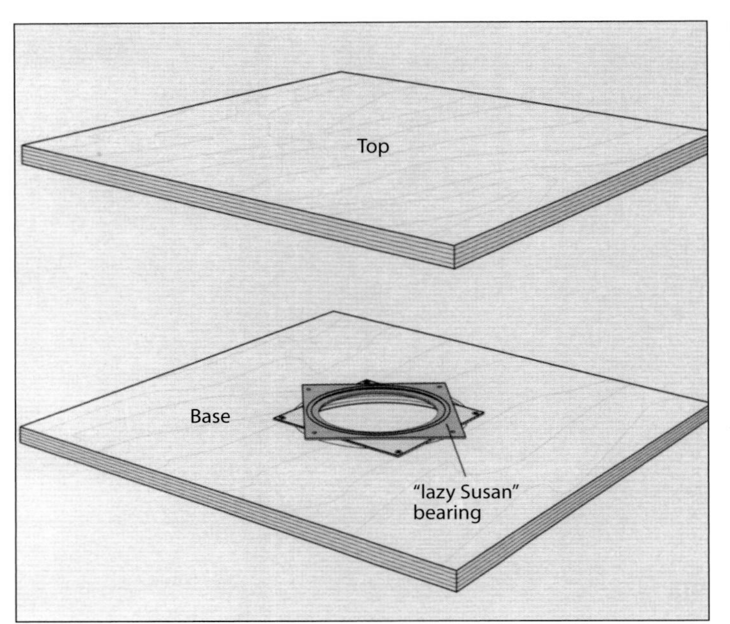

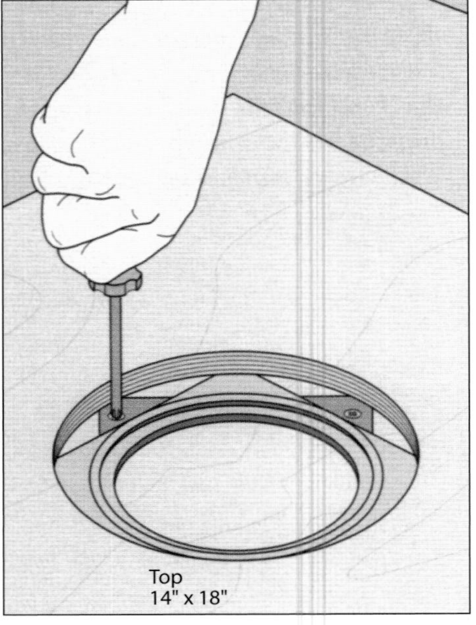

Building the turntable

Consisting of a base and top cut from ¾-inch plywood with a "lazy Susan" bearing fastened in between, the turntable shown above allows a piece of furniture to be rotated as it is being sprayed with a finish. Cut the base and top slightly larger than the base of the piece of furniture to be finished. Cut a hole in the center of the base to allow access to the screw holes for attaching the upper bearing to the top once the lower bearing is secured to the base. First attach the lower bearing to the base with screws. To fasten the jig top, set the base on top of it with the bearing sandwiched between the two pieces and flip them upside down. With the edges of the pieces flush, rotate the bearing so the remaining screw holes are exposed, then screw the upper bearing to the top *(above, right)*.

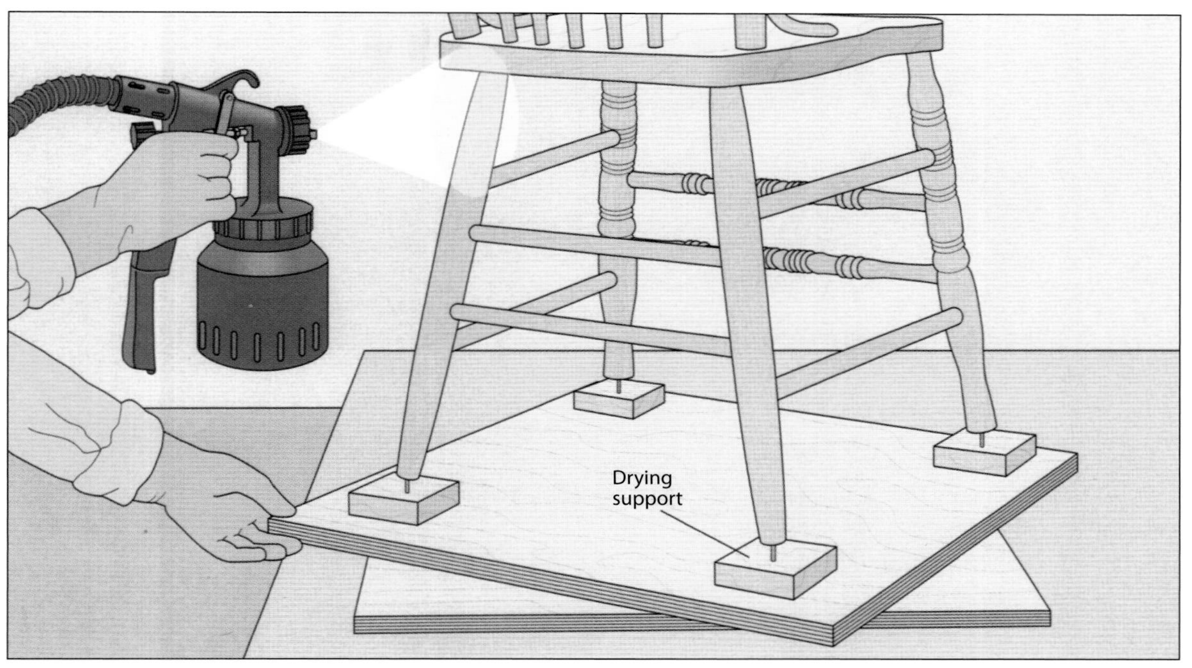

Using the turntable
Make four small drying supports. Set the workpiece on the tips of the nails, then slowly rotate the turntable with one hand while operating a spray gun with the other *(above)*.

Bench Dog Lamp Support

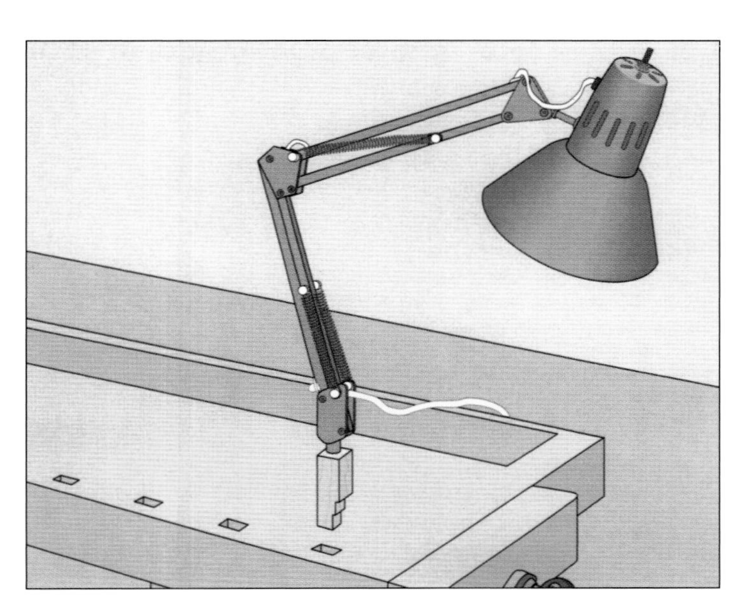

A movable light for a workbench
A desk lamp attached to a bench dog as shown at left will enable you to position the light at any of the dog holes along the bench. To make the jig, bore a hole the same diameter as the shaft of the lamp into the head of a wooden bench dog *(page 38)*.

Safety

For most woodworkers, the home workshop is a peaceful refuge, where craft gives shape to creative ideas. It is also the place where accidents may occur, owing to the very nature of the activity. But the likelihood of mishap can be reduced by a few simple precautions. First, an informed woodworker is a safe woodworker. Read the owner's manuals supplied with all your tools. Before starting a job, make sure you know how to use the safety accessories that are designed to protect you from injury while working with a tool.

Most accidents are the result of carelessness or inattention—failure to use a safety guard when cutting a board on a table saw, face jointing stock with bare hands (rather than with a push block), or using a router without safety goggles. Refer to the safety tips on page 132 for ways of avoiding some of the more common accidents in the shop.

Although the big stationary machines receive most of the attention from safety-conscious woodworkers, there are other potential sources of danger that, though less apparent, cannot be ignored. Many finishing products, particularly those containing solvents, can be toxic, although their effects may only become apparent after years of prolonged exposure. Certain species of wood can cause allergic or toxic reactions in some people. Page 133 presents information on choosing safe finishing products and on the possible health effects of some wood species. Safety goggles, rubber gloves, and a rubber apron are good standard attire for any finishing job, especially if you are spraying a finish or mixing and applying caustic chemicals.

Fire is another shop hazard. Smoke detectors are an invaluable defence, providing valuable time for you to control the blaze *(page 134)*. Keep a fire extinguisher rated ABC in your shop and know how to use it. One of the leading causes of fire is improper wiring. Whether you are building a shop from scratch or revamping an existing space, electrical safety should be a priority *(page 135)*.

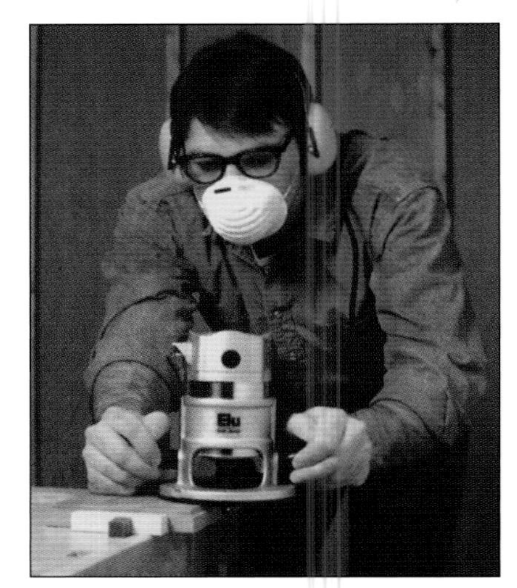

Personal safety gear is one insurance against injury. Here, a woodworker routs a groove in a drawer front, wearing safety glasses, a dust mask, and ear muffs.

No shop should be without the personal safety gear illustrated on page 136. You can easily make some safety devices, such as push sticks, push blocks, and featherboards *(page 138)*. But do not become complacent about the security they will provide. All the safety equipment in the world cannot make a shop accident-free. Safety is foremost a matter of attitude—a confidence in using the machines combined with a healthy respect for the power these tools wield.

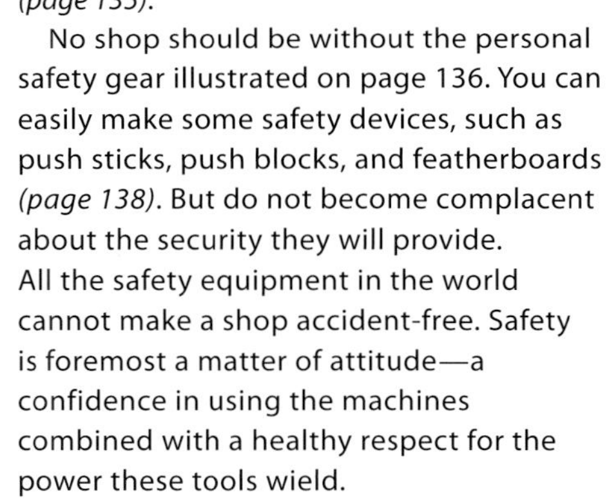

Even with the best efforts at prevention, accidents still occur. Bits may break, boards split, shavings fly and all too often find a victim. Being prepared and taking prompt action can help minimize further damage. Take a first-aid course, keep a well-stocked first-aid kit on hand in the shop *(page 144)* and be ready to administer medical aid when necessary.

There are many safety devices that can minimize the risk of using power tools. The table saw in this photo features a plastic shield that covers the blade; the splitter and the anti-kickback pawl protect against binding and kickback. A hold-down device presses the workpiece flat on the table and firmly against the fence. A push stick allows the woodworker to feed the stock into the blade while keeping fingers well away from the cutting edge.

Accident Prevention

Safety Tips

General

- Make sure workshop lighting and ventilation are adequate.
- Keep children, onlookers, and pets away from the work area.
- Concentrate on the job; do not rush or take shortcuts. Never work when you are tired, stressed, or have been drinking alcohol or using medications that induce drowsiness.
- Find a comfortable stance; avoid over-reaching.
- Keep your work area clean and tidy; clutter can lead to accidents.

Hand Tools

- Use the appropriate tool for the job; do not try to make a tool do something for which it was not designed.
- When possible, cut away from yourself rather than toward your body.
- Keep tools clean and sharp.

Power Tools

- Wear appropriate safety gear: safety glasses or face shield and hearing protection. If there is no dust collection system, wear a dust mask. For allergenic woods, such as ebony, use a respirator.
- Read your owner's manual carefully before operating any tool.
- Tie back long hair and avoid loose-fitting clothing. Remove rings and other jewelry that can catch in moving parts.
- Unplug a tool before performing setup or installation operations.
- Whenever possible, clamp down the workpiece, leaving both hands free to perform an operation.
- Keep your hands well away from a turning blade or bit.
- Turn off a tool if it produces an unfamiliar vibration or noise; have the tool serviced before resuming operations.
- Do not use a tool if any part of it is worn or damaged.

Finishing

- Do not eat, drink, or smoke when using finishing products.
- Avoid exposure to organic solvents if you are pregnant or breast-feeding.
- Install at least one smoke detector on the ceiling of your shop above potential fire hazards; keep a fully charged ABC fire extinguisher nearby.
- Never store solvents or chemicals in unmarked containers. Chemical solutions should always be stored in dark glass jars to shield them from light, which may change their composition.
- Store finishing products in a locked cabinet.
- To prevent eye injury, wear safety goggles, and don rubber gloves when working with caustic or toxic finishing products.
- Do not flush used solvents down the drain. Consult the Yellow Pages to find out who handles chemical disposal in your area, or check with your local fire department.

Shop Tip

Disabling a power tool

To prevent unauthorized use of a power tool, slip the bolt of a mini-padlock through one of the tines in the power cord plug. The lock will make it impossible to plug in the tool. If you are using a keyed lock, store the keys out of the way in a cupboard or drawer that can be locked.

Working with Safe Finishes

Although a number of high-quality water-based finishes have become available recently, solvent-based finishing products are still widely used, and considered superior for some applications. Thus woodworkers must learn to protect themselves against the health hazards associated with organic solvents. Organic solvents can have a number of health effects. Short-term use can result in ailments ranging from headaches and nausea to skin and eye irritation. With extended use, many solvents are known to damage the central nervous system or respiratory tract. Some glycol ethers are suspected of causing birth defects, while other solvents, like methylene chloride, have been linked with cardiac arrest.

Solvents can be absorbed into the bloodstream in a number of ways: after being inhaled, or ingested along with food left in the shop, absorbed through the skin, or swallowed when vapors settle in saliva. Most solvent-based finishes are unlikely to cause harm when used occasionally, and are only poisonous if swallowed. But you still need to be aware of the combination and concentration of organic solvents in a particular finish if you plan to use the product in large quantities or over an extended period of time. The chart below lists the solvents contained in a variety of finishing products and assesses the relative toxicity of each one. Be sure to choose the safest product for the job at hand.

Toxic Solvents

Finishing Product	Solvent
Wood filler (paste and liquid)	Petroleum naphtha,* mineral spirits,* acetone,** methyl ethyl ketone,** methyl isopropanol,** isobutyl ketone***
Stains (aniline, wiping, NGR, gel and glazing stains; color pigments)	Ethanol,* mineral spirits,* toluene,*** xylene,*** methanol,*** glycol ethers***
Shellacs (white and orange)	Ethanol,* methanol***
Lacquers (spray and brush, sanding sealers)	Acetone,** methyl ethyl ketone,** isopropanol,** methanol,*** xylene,*** glycol ethers***
Lacquer thinner	Acetone,** methyl ethyl ketone,** isopropanol,** glycol ethers,*** toluene***
Rubbing oils (Danish oil, antique oil)	VM&P naphtha,* turpentine,** toluene***
Drying oils (boiled linseed oil, polymerized tung oil)	Mineral spirits,* turpentine**
Varnishes (tung oil varnish, spar varnish, varnish stain)	Mineral spirits,* VM&P naphtha*
Polyurethanes (poly varnish, urethane stains)	Mineral spirits,* toluene***
Lacquer/varnish removers	Acetone,** xylene,*** methanol,*** methyl isobutyl ketone,*** toluene***
Waxes (paste wax, furniture wax)	Petroleum naphtha,* turpentine**
	*** Safest product ** Mildly hazardous product *** Product to be avoided if possible**

Safety

Back to Basics

Fire Safety

Considering the number of flammable materials and potential ignition sources in a woodworking shop, fire prevention should be one of your foremost safety concerns. Sawdust, wood, paint, and thinners tend to accumulate; often they are near tools that produce sparks and heat. The combination can prove volatile: When vaporized in a small enough concentration of air, a small quantity of lacquer thinner, for example, can be ignited by a spark from a tool and cause a life-threatening explosion.

The first step in fire safety is prevention. All finishing products and solvents, for example, should be stored away from heat sources in airtight glass or metal containers, preferably in a fireproof cabinet. Hang rags soaked with flammable chemicals to dry outdoors, or soak them in water and store them in sealed metal containers. When working with finishing products, keep windows open and the shop well ventilated.

Be prepared to deal with a fire effectively. Install a smoke detector on the shop ceiling or a wall, and keep an ABC fire extinguisher nearby. Design a fire evacuation plan that maps out two possible escape routes from each room of the building in which the shop is located. If the fire involves an electric tool, a power cord, or an electrical outlet, shut off the power. Call the fire department immediately, inform them of the nature of the fire, and try to extinguish the blaze yourself. But if the flames cannot be contained, or the fire is coming from inside a wall or ceiling, evacuate the building.

Controlling a fire

To extinguish a small, contained fire, use an ABC-rated dry-chemical fire extinguisher, which is effective against all three major classes of fires: burning wood or other combustibles (Class A), oil- or grease-fed flames (Class B), and electrical blazes (Class C). Position yourself a safe distance from the fire with your back to the nearest exit. Holding the extinguisher upright, pull the lock pin out of the handle *(inset)* and aim the nozzle at the base of the flames. Squeeze the handle and spray in a quick, side-to-side motion *(right)* until the fire is out. Watch for "flashback," or rekindling, and be prepared to spray again. If the fire spreads, leave the building. Dispose of burned waste following the advice of the fire department. After use, have the extinguisher professionally recharged; replace it if it is non-rechargeable.

Electrical Safety

Electricity plays a major role in the modern woodworking shop, powering machines and tools, lighting fixtures and lamps, and heating systems. Electricity is so commonplace that it is all too easy to forget its potential for danger. An electrical shock, even one that can hardly be felt, can be deadly. For this reason, the electrical system is strictly regulated by codes and standards designed to protect you from fire and shock.

Living safely with electricity also requires following basic precautions designed to prevent mishaps. Inspect plugs for cracks and power cords for fraying, and replace any worn or damaged part before using a tool. Never replace a blown fuse with one of a higher amperage. Do not plug a three-prong plug into a two-slot outlet by removing the grounding prong from a three-prong plug. Instead, replace the outlet with a GFCI *(below)*.

Before undertaking a repair, shut off the power at the service panel. To work on the system, wear rubber gloves and, where possible, use only one hand, keeping your free hand behind your back.

Plugging in Safely

Using GFCI outlets

The U.S. National Electrical Code requires that any new outlet in a garage or unfinished basement must be protected by a ground-fault circuit interrupter (GFCI). A GFCI protects a circuit—and you—by monitoring the flow of electricity passing through it and tripping instantly when it detects a leak to ground. If you need to replace an outlet in your shop, install a GFCI, such as the one shown above, following the manufacturer's directions, or have a qualified electrician do the work. Test the outlet once every month by pushing the TEST button; the RESET button should pop out. If it does not, have the outlet serviced. To reactivate the outlet, press the RESET button.

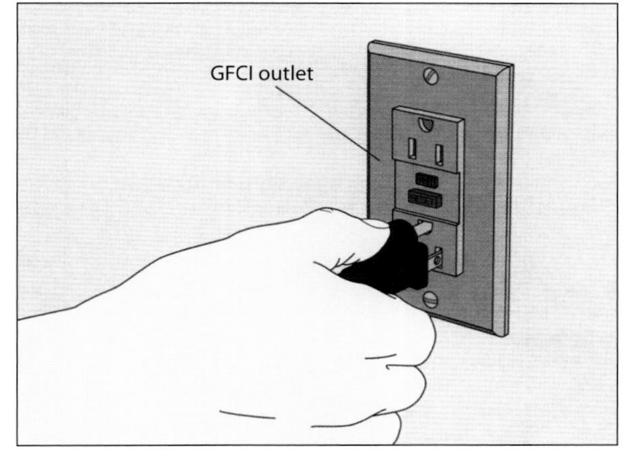

Amperage Rating of Tool	Minimum Gauge for Different Length Cords		
	25'	50'	75'
0–2.0	18	18	18
2.1–3.4	18	18	18
3.5–5.0	18	18	16
5.1–7.0	18	16	14
7.1–12.0	18	14	12
12.1–16.0	16	12	10

Minimum Wire Gauge for Extension Cords

Choosing a wire with the proper gauge

Using an extension cord with the wrong gauge can cause a drop in line voltage, resulting in loss of power, excessive heat, and tool burnout. Refer to the chart at left to determine the minimum wire gauge for the tool and task at hand. If, for instance, your tool has a 7-amp motor and you are using a 75-foot extension cord, the minimum gauge should be 14. Choose only round-jacketed extension cords listed by Underwriters Laboratory (UL).

Personal Safety Gear

The personal safety equipment shown below can go a long way toward shielding you from most dangers in the workshop. But carrying an inventory of safety gear is not enough; the items must be properly used to protect you from injury.

The need for some items may not be readily apparent, although the dangers are very real. Few woodworkers need to be reminded of the cutting power of a spinning saw blade or jointer cutterhead. Less well known are the long-term effects of being exposed to the sound generated by power tools. The chart on the next page lists a variety of power tools along with their approximate noise levels in decibels. The chart also indicates the longest recommended time that an unprotected person can be exposed to various levels before risking permanent hearing loss.

Remember, too, that even short-term exposure to some noise, while it may not lead to hearing loss, can dull the senses and cause a woodworker's alertness to flag—a setup for an accident.

A Panoply of Safety Equipment

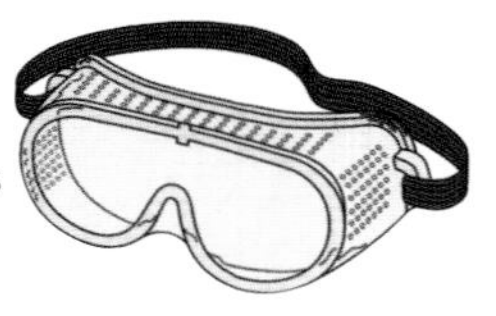

Safety goggles
Flexible, molded plastic goggles protect eyes. Type with perforated vent holes shields against impact injury and sawdust; type with baffled vents protects against chemical splashes; nonvented goggles also available.

Face shield
Clear plastic shield protects against flying debris and splashes; features adjustable head gear.

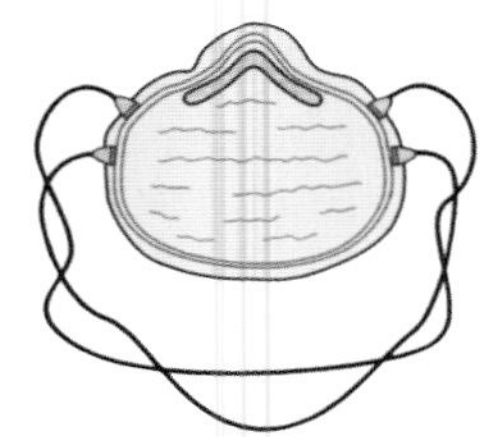

Disposable dust mask
Fits over nose and mouth for one-time-use protection against inhalation of dust or mist; features a cotton or fiber shield with an adjustable head strap and a metal nose clip.

Rubber gloves
Household rubber gloves or disposable vinyl gloves protect against mild chemicals or finishes; neoprene rubber gloves shield skin from caustic finishing products.

Dual-cartridge respirator
Protects against fumes when working with chemicals or spraying a finish. Interchangeable filters and chemical cartridges shield against specific hazards; filter prevents inhalation of dust. Cartridges purify air and expel toxins through exhalation valve.

Reusable dust mask
Features a neoprene rubber or soft plastic frame with an adjustable head strap and a replaceable cotton fiber or gauze filter; protects against dust and mist.

Ear muffs
Cushioned muffs with adjustable plastic head strap protect hearing against high-intensity noise from power tools.

Ear plugs with neckband
Detachable foam-rubber plugs compressed and inserted into ear canals provide hearing protection from high-intensity power tool noise; plastic neckband fits around neck.

Work gloves
For handling rough lumber; typically features leather or thick fabric palms and fingertips with elasticized or knitted wrists.

Safety glasses
Standard plastic frames fitted with shatterproof lenses protect eyes from flying wood chips and other debris; typically feature side shields.

Noise Levels Produced by Power Tools

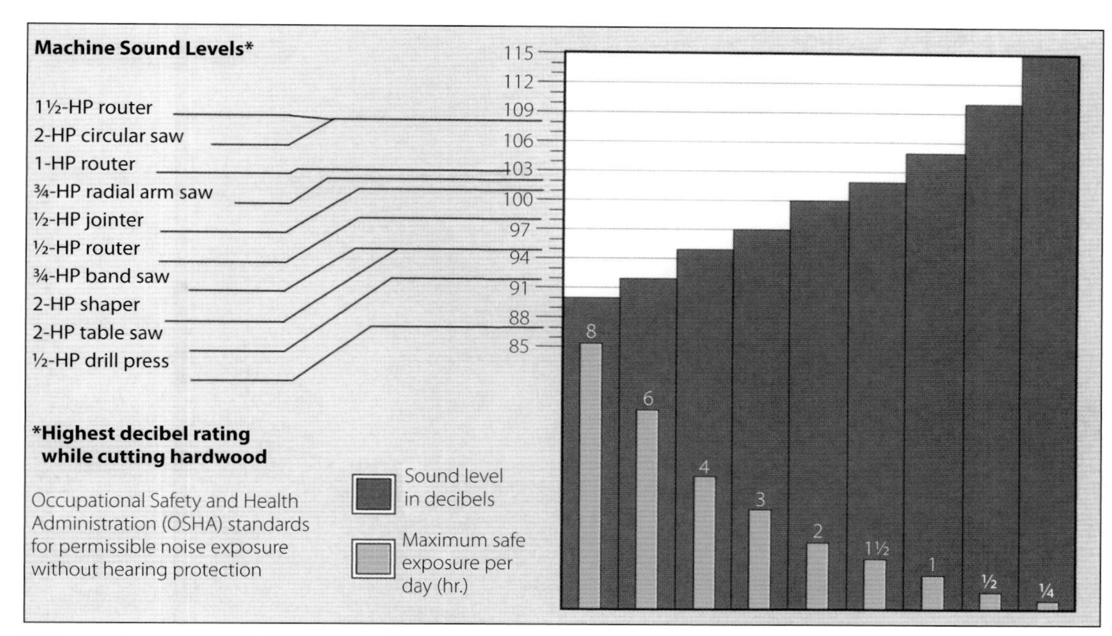

While a ½-horsepower drill press is unlikely to damage your hearing—unless you run the machine all day long—unprotected exposure to the noise produced by a 1½-horsepower router can be dangerous after only 30 minutes. The above chart shows approximate noise levels produced by a variety of power tools. Keep in mind that tools with dull cutters or blades generate more noise than those with well-sharpened cutting edges.

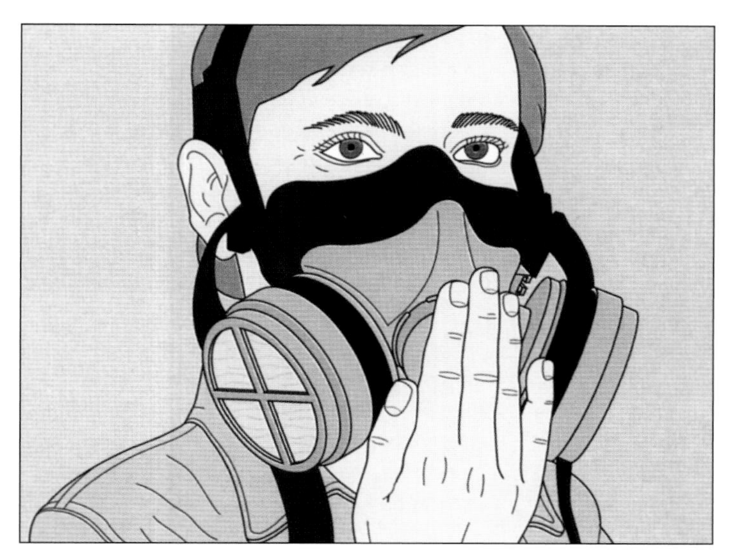

Testing a Respirator

Checking for air leaks

A respirator is only as good as its seal against your face. No seal, no protection. To test your respirator, place it over your face, setting the top strap over the crown of your head. Adjust the side straps for a snug fit. To test the respirator, cover the outlet valve with your hand and breathe out gently *(left)*. There should be no air leakage around the facepiece. If air leaks out of the respirator, readjust the straps for a tighter fit. Replace the facepiece when necessary following the manufacturer's instructions, or replace the respirator. Use the appropriate filters for the job at hand. (If you have a beard, use a full-face mask with forced-air ventilation.)

Safety Devices

Push Sticks and Push Blocks

Making push sticks and push blocks
Push sticks and push blocks for feeding stock across the table of a stationary power tool can be made using ¾-inch plywood or solid stock. No one shape is ideal; a well-designed push stick should be comfortable to use and suitable for the machine and task at hand. For most cuts on a table saw, design a push stick with a 45° angle between the handle and the base *(right, top)*. Reduce the handle angle for use with the radial arm saw. The notch on the bottom edge must be deep enough to support the workpiece, but shallow enough not to contact the saw table. The long base of a rectangular push stick *(right, middle)* enables you to apply downward pressure on a workpiece. For surfacing the face of a board on a jointer, the long, wide base of a push block *(right, bottom)* is ideal. It features a lip glued to the underside of the base, flush with one end. Screw the handle to the top, positioning it so the back is even with the end of the base.

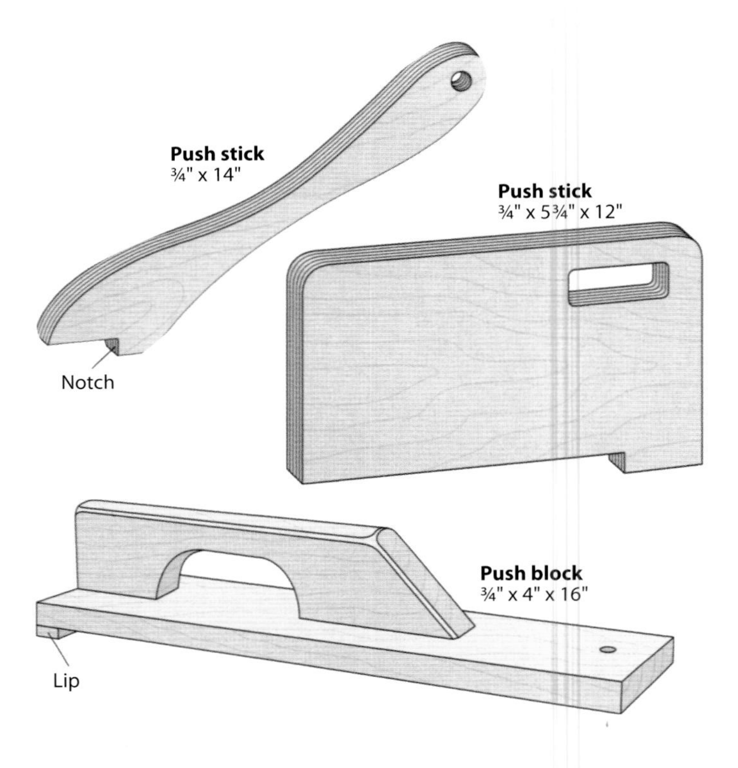

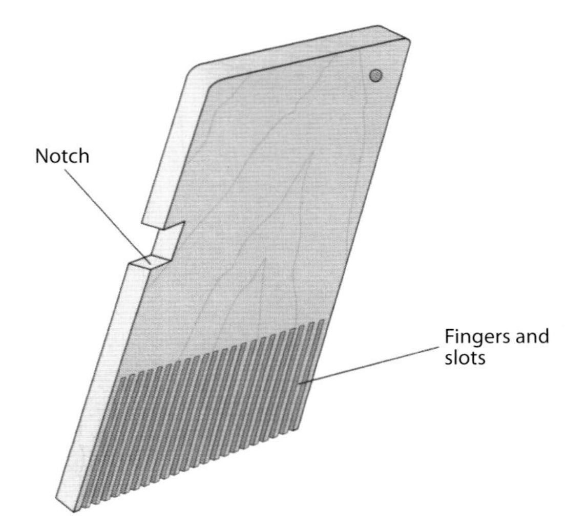

A Standard Featherboard

Making a standard featherboard
Featherboards serve as anti-kickback devices, since the fingers allow the workpiece to move in only one direction—toward a stationary tool's bit or blade. To make a featherboard like the one shown at left, cut a 30° to 45° miter at one end of a ¾-inch-thick, 3- to 4-inch-wide board; the length of the jig can be varied to suit the work you plan to do. Mark a parallel line about 5 inches from the mitered end and cut a series of slots to the marked line on the band saw, spacing the kerfs about ⅛ inch apart to create a row of sturdy but pliable fingers. Finally, cut a notch out of one edge of the featherboard to accommodate a support board.

Using standard featherboards on the table saw

Clamp one featherboard to the fence above the blade, and place a longer one halfway between the blade and the front of the table. Clamp a support board in the notch perpendicular to the horizontal featherboard to prevent it from creeping out of place during the cut. For the operation shown at left, feed the workpiece into the blade until your trailing fingers reach the featherboards. Then use a push stick to finish the cut, or move to the back of the table with the saw still running and pull the workpiece past the blade.

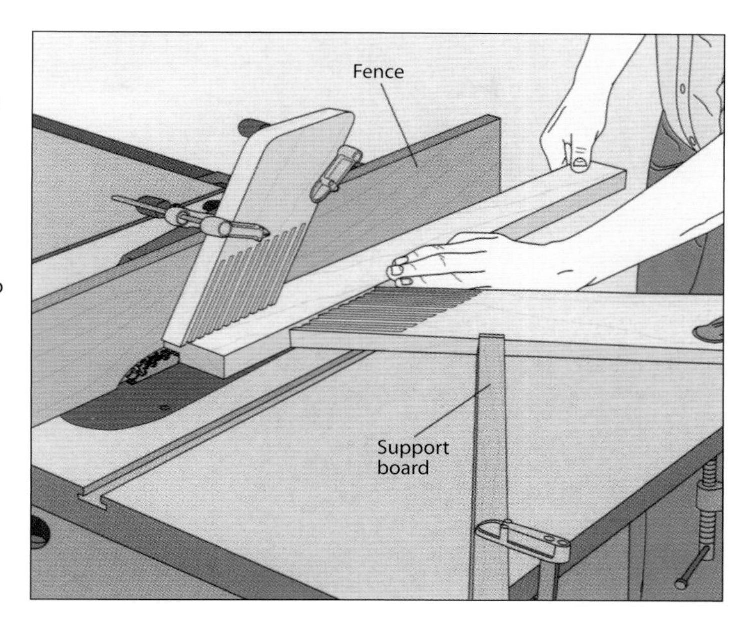

Push sticks and featherboards make an operation like ripping on the table saw much safer by keeping your hands well away from the blade. The push stick is used to feed the stock and keep it flat on the table, while the featherboard presses the workpiece against the fence. The featherboard shown in the photo is secured to the table with special hardware rather than with clamps. A clamping bar in the miter slot features two screws that can be tightened, causing the bar to expand and lock tightly in the slot.

A Beveled Featherboard

Ripping stock with a beveled featherboard
A featherboard clamped to the fence of a table saw, as shown on the previous page, can get in the way of a push stick during a rip cut. A featherboard with a beveled end will press a workpiece against both the fence and saw table, eliminating the need to clamp a featherboard to the fence *(right)*. Make the device as you would a standard featherboard *(page 138)*, but cut a 45° bevel on its leading end before cutting the fingers and slots. Also make sure that the featherboard is thicker than the stock you are ripping *(inset)*.

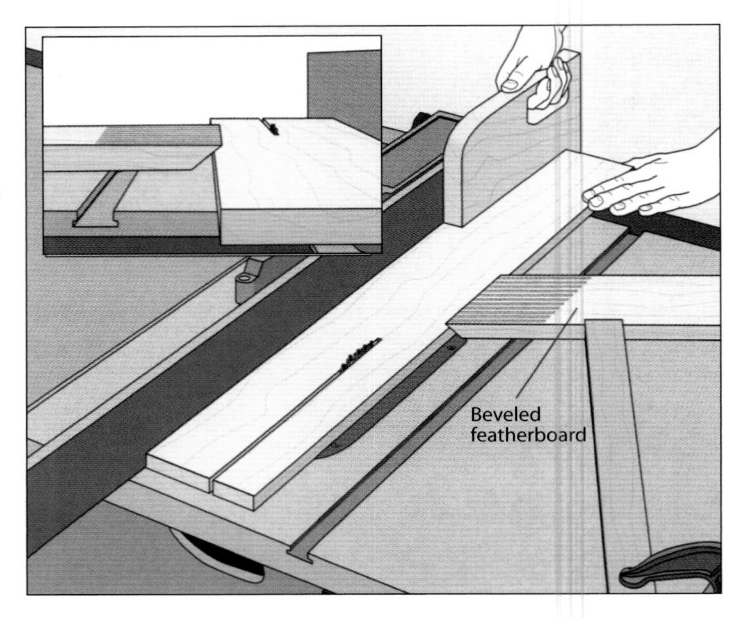

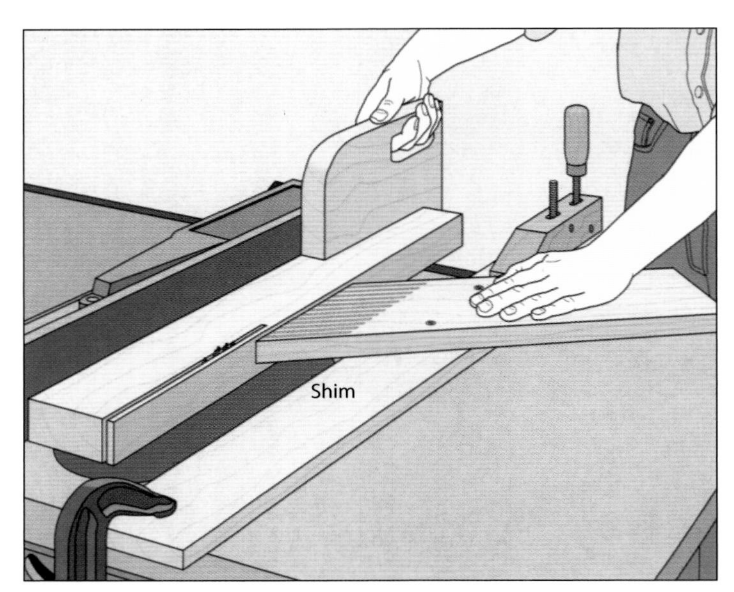

A Shimmed Featherboard

Shimming a featherboard
When working with thick stock or running a board on edge across a saw table, a featherboard clamped directly to the table may apply pressure too low on the workpiece, causing it to tilt away from the fence. To apply pressure closer to the middle of the stock, screw the featherboard to a shim and then clamp the shim to the table *(left)*.

Push Block

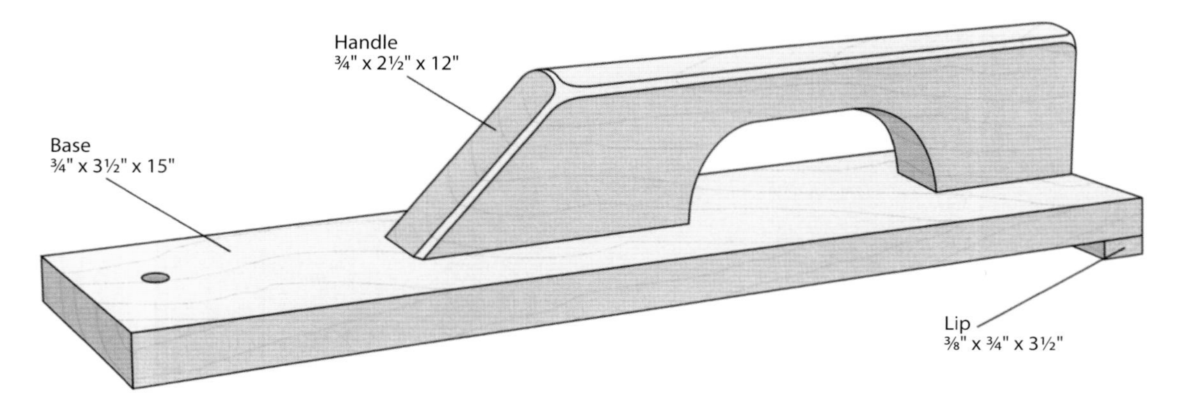

A Push Block For Face Jointing

The long, wide base of the push block shown above is ideal for surfacing the face of a board on a jointer. Although push blocks for such jobs are available commercially, you can easily fashion your own. Refer to the illustration for suggested dimensions, but tailor the design to suit your own needs.

Cut the pieces to size, then glue the lip to the underside of the base, flush with one end. Screw the handle to the top, positioning it so the back is even with the end of the base. Drive the screws from the underside of the base; be sure to countersink the fasteners to avoid marring the workpiece when you feed it across the jointer knives. Bore a hole near the front end of the base so you can hang the push block on the wall when it is not in use.

To use the push block, set the workpiece on the jointer's infeed table a few inches from the knives, butting its edge against the fence. Then lay the push block squarely on top of the stock, centered between its sides, with the lip over the trailing end of the workpiece. With your leading hand on the front end of the stock and your thumb braced against the push block, slowly feed the workpiece across the knives *(left)*. (For stock thinner than ¾ inch, use only the push block.) Apply downward pressure to keep the stock flat on the tables and lateral pressure to keep it butted against the fence.

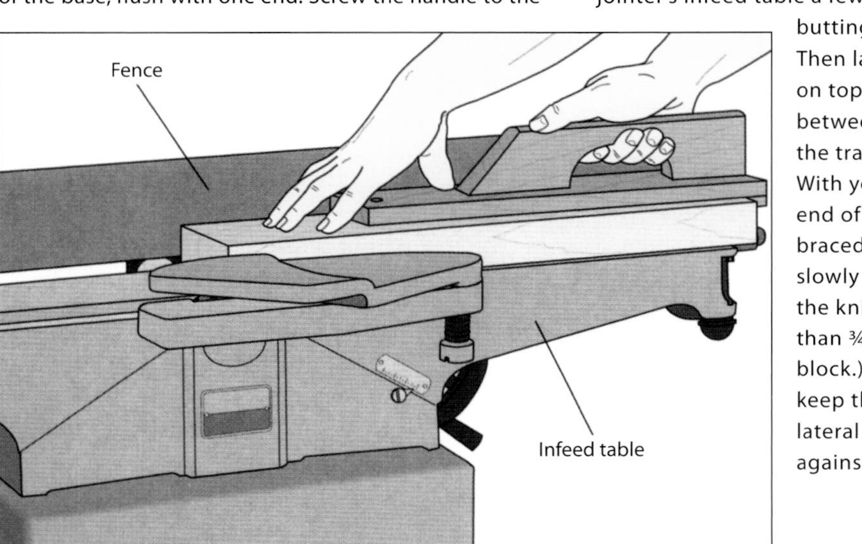

Planing Short and Thin Stock

Using runner guides to plane short stock
Feeding short boards through a thickness planer can cause sniping and kickback. To hold short stock steady as it enters and exits the planer, glue two solid wood scrap runners to the edges of your workpiece. Make sure the runners are the same thickness as the workpiece and extend several inches beyond both ends. Feed the workpiece into the planer *(right)*, making a series of light cuts until you have reached the desired thickness. Then cut off the runners.

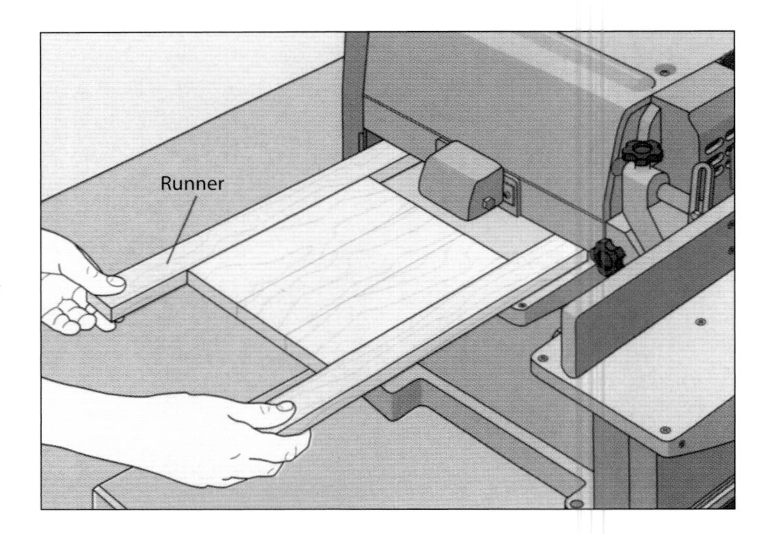

Using a planing jig for thin stock
Thickness planing stock thinner than ¼ inch often causes chatter and splintering of the workpiece. To avoid these problems make thin stock "thicker" with this jig. To make it, simply glue two beveled cleats to either end of a board that is slightly longer than your workpiece *(inset)*. To make the cleats, cut a 45° bevel across the middle of a board approximately the same thickness as the workpiece. Next, bevel the ends of the workpiece. Set the stock on a backup board, position the cleats flush against the workpiece so the bevel cuts are in contact, and glue the cleats in place to the backup board. Run the jig and workpiece through the planer, making several light passes down to the desired thickness (right), then crosscut the ends of the workpiece square.

An Auxiliary Switch for the Table Saw

Installing an overhead switch
Switching on a table saw while balancing a large panel on the table can prove difficult. The addition of an overhead switch will enable you to start the saw when the main switch is out of reach *(right)*. Locate the new switch so you can reach it comfortably with a 4-by-8 panel on the saw table just in front of the blade; screw a triangular bracket to the ceiling and attach the switch to the bracket at a suitable height. Run a length of non-metallic sheathed 12-gauge cable from the switch along the ceiling, down the wall, and across the floor to your saw. Have a licensed electrician wire the switch to the saw so that both it and the original switch are able to start or stop the machine; never disconnect the switch on the saw itself.

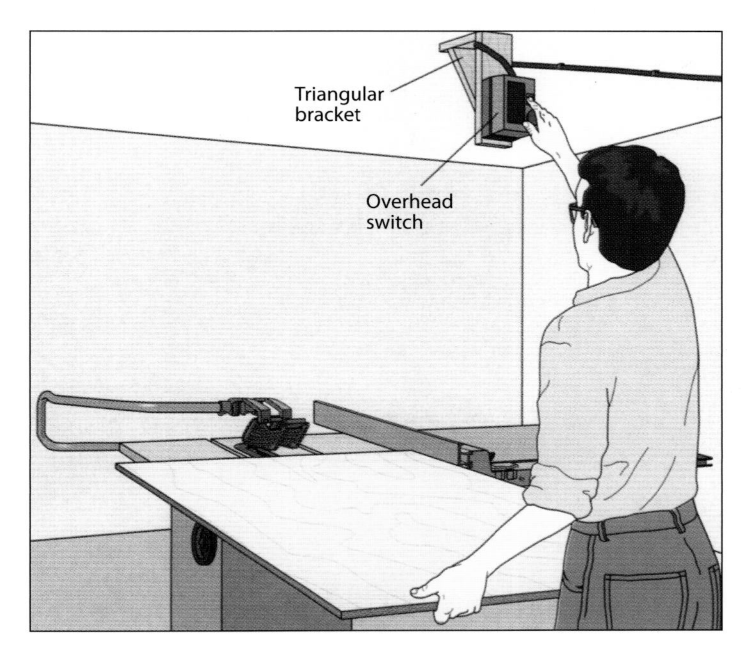

Shop Tip

A safe attention-getter
If the door to the shop is outside your field of vision when you are at a machine, there is the risk that someone might enter the shop, tap you on the shoulder, and startle you. To avoid accident-causing surprises, mount a light bulb at eye level near the tool and wire the switch to the door frame so that the bulb lights when the door is opened. Wiring another bulb to the bell circuit of the telephone can solve the problem of missing phone calls: Each time the bell rings, the bulb will light.

First Aid

Most woodworking accidents arise from the improper use of tools and safety guards, unsafe work habits, and mishandling hazardous materials. Take the time to set up properly for a job, gathering together the tools, equipment, and materials you need. Always use the appropriate safety gear. Work methodically; never hurry through a job. Be especially careful—or stop working—if you are fatigued.

Accidents can befall even the most careful woodworker. Boards split, blades nick, and liquids splash. Many finishing products contain chemicals that emit toxic fumes, causing dizziness or nausea. Store a first-aid kit, stocked with the basic supplies shown below, in an easily accessible spot in your shop. Keep emergency telephone numbers handy. Techniques for handling some common shop mishaps are shown on the following pages.

First-Aid Supplies

Adhesive bandages
Sterile gauze dressings with adhesive strips for protecting scratches or minor cuts. Available in a wide variety of sizes and shapes: square, rectangular, round, butterfly, and fingertip.

Medical tape
Secures gauze dressings, gauze roller bandages or eye pads; hypo-allergenic for sensitive skin. Available in lengths of 2½ to 10 yards and in widths of ½ to 3 inches.

Gauze roller bandage
Sterile roll secures gauze dressings; fastened with medical tape or safety pin, or by knotting. Available in lengths of 5 to 10 yards and widths of 1 to 4 inches.

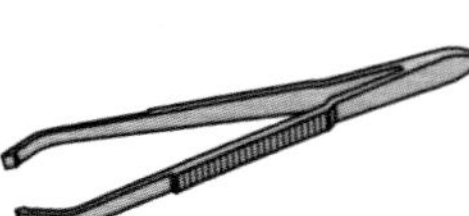

Tweezers
Extract splinters or other small objects lodged in skin. Made of stainless steel in a variety of shapes and sizes; flat-tipped type 4½ inches long is common.

Ipecac syrup
For inducing vomiting in a poisoning victim.
Caution: Administer only If advised by a physician or poison control center.

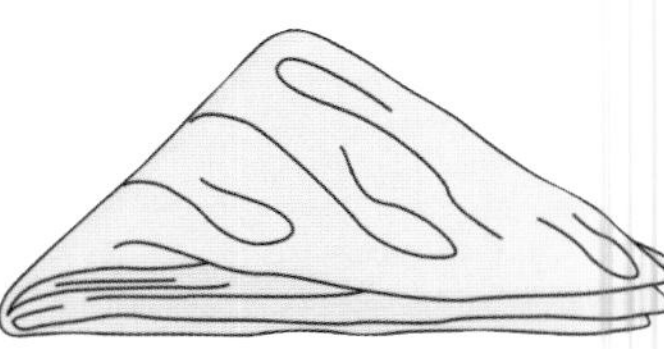

Triangular bandage
Multipurpose cotton bandage can be folded to make sling, pad, or bandage; measures 55 inches across base and 36 to 40 inches along each side.

Rubbing alcohol
Also known as isopropyl alcohol; sterilizes tweezers and other first-aid equipment.

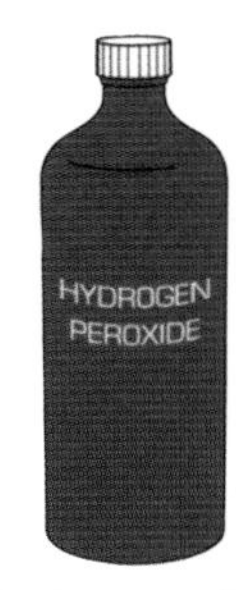

Hydrogen peroxide
For cleaning wounds before applying adhesive bandages, gauze dressings, or gauze roller bandages; commonly available in 3% solution.

Gauze dressing
Sterile pad for covering a wound; secured with medical tape or gauze roller bandage. Available in sizes of 2-by-2, 3-by-3 and 4-by-4 inches.

Eye irrigator
Filled with water and used to flush foreign particles from eye.

Eye pads
Sterile pads taped over eyes to protect them and prevent movement; self-adhesive patches also available.

Providing Minor First Aid

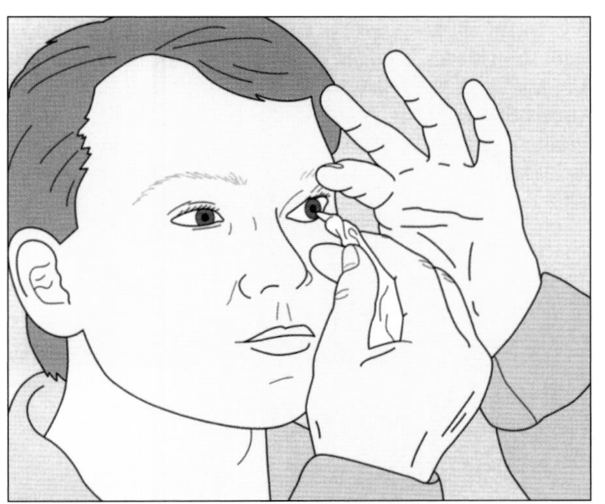

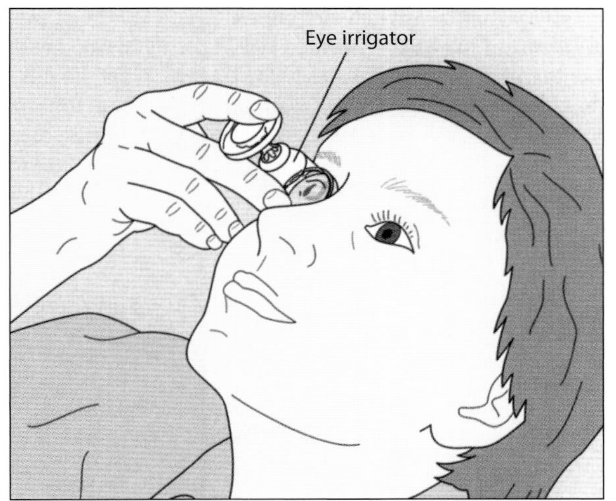

Clearing a particle from the eye
Hold your affected eye open with the forefinger and thumb of one hand. Slowly rotate your eye, if necessary, to help expose the particle. Gently wipe away the particle using the twisted end of a tissue moistened with water *(above, left)*. Or, fill an eye irrigator with cool water and use it to flush out the particle. Lean forward with both eyes closed and press the rim of the irrigator against the affected eye, and tilt back your head. Open your eyes *(above, right)* and blink several times to flush out the particle. If you cannot remove the particle, seek medical help immediately. **Caution:** Do not remove a particle that is on the cornea, is embedded, or has adhered to the eye.

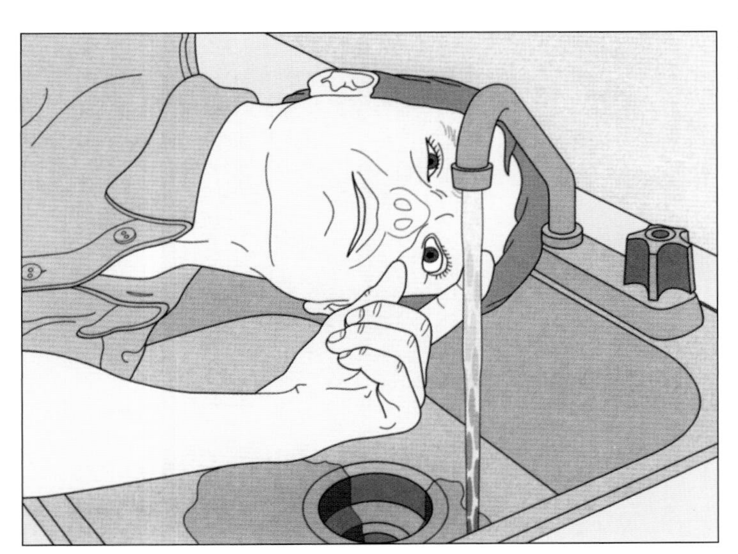

Flushing a chemical from the eye
Holding the eyelids of the affected eye apart, flush the eye thoroughly for at least 15 minutes under a gentle flow of cool water from a faucet *(right)* or pitcher; tilt your head to one side to prevent the chemical from being washed into the uninjured eye. If you are outdoors, flush the eye using a garden hose. Gently cover both eyes with eye pads or sterile gauze dressings and seek medical help immediately.

Pulling out a splinter

Wash the skin around the splinter with soap and water. (A metal splinter, even if you are able to remove it, may require treatment for tetanus; seek medical help.) To remove the splinter, sterilize a needle and tweezers with rubbing alcohol. Ease the end of the splinter out from under the skin using the needle, then pull it out with the tweezers *(right)*. Clean the skin again with soap and water. If the splinter cannot be removed, seek medical attention.

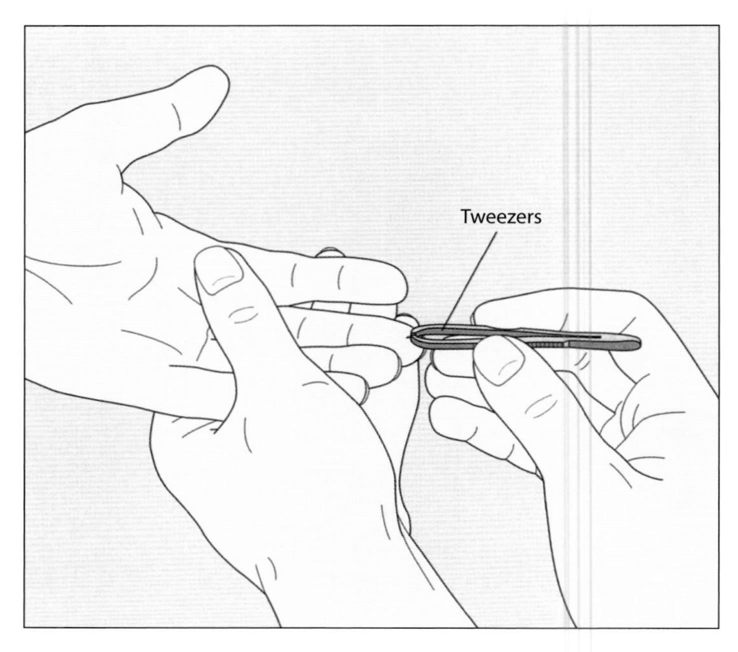

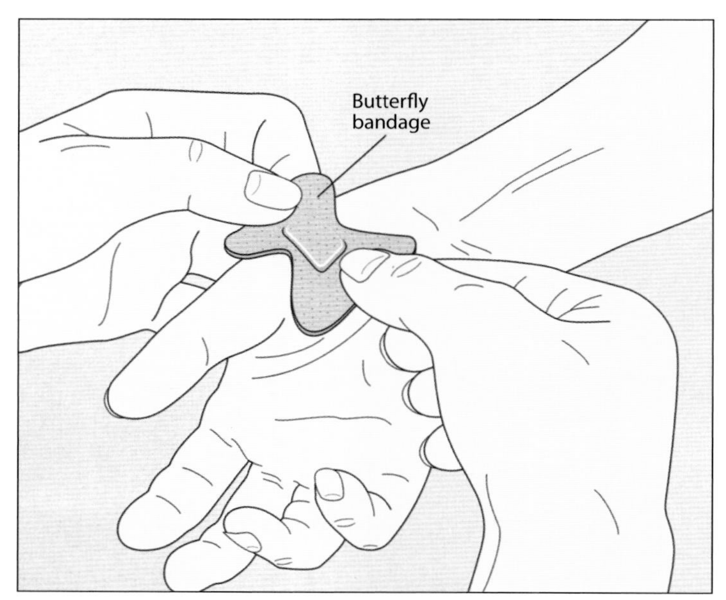

Treating a cut

Wrap the wound in a clean cloth and apply direct pressure with your hand to stop any bleeding; keep the wound elevated. If the cloth becomes blood-soaked, wrap another cloth over it. If bleeding persists or the wound is deep or gaping, seek medical help. Otherwise, wash the wound with soap and water, then bandage it; for a narrow, shallow wound, draw its edges closed with a butterfly bandage *(left)*.

Controlling Bleeding

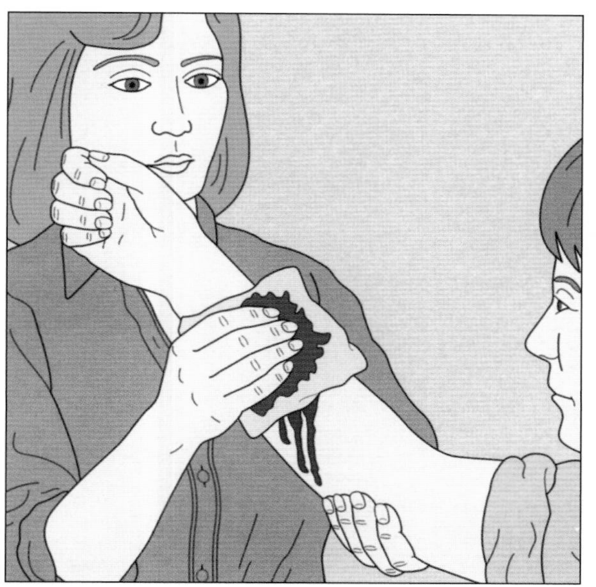

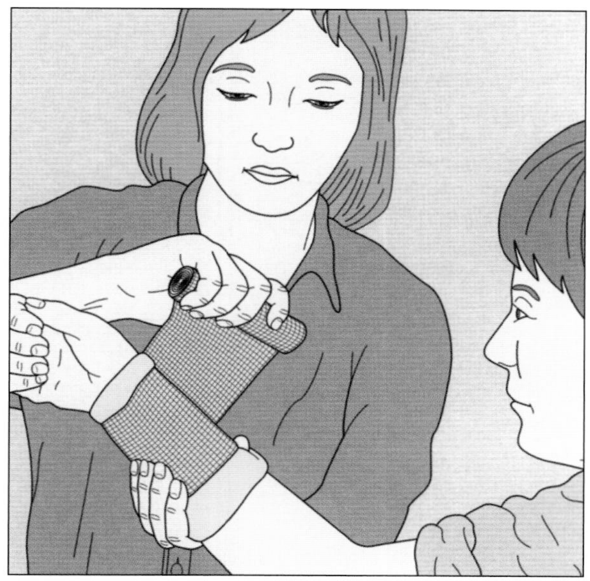

Applying direct pressure to stop bleeding

To help stop profuse or rapid bleeding, apply direct pressure to the wound with a gauze dressing or a clean cloth and, if possible, elevate the injury *(above, left)*. Direct pressure should stop the flow of blood and allow it to clot. If the dressing becomes blood-soaked, add another over the first one; avoid lifting the dressing to inspect the wound. It will be easier to maintain steady pressure if you wrap the wound with a roller bandage *(above, right)* for added direct pressure. If you cannot stop the bleeding, call for medical help.

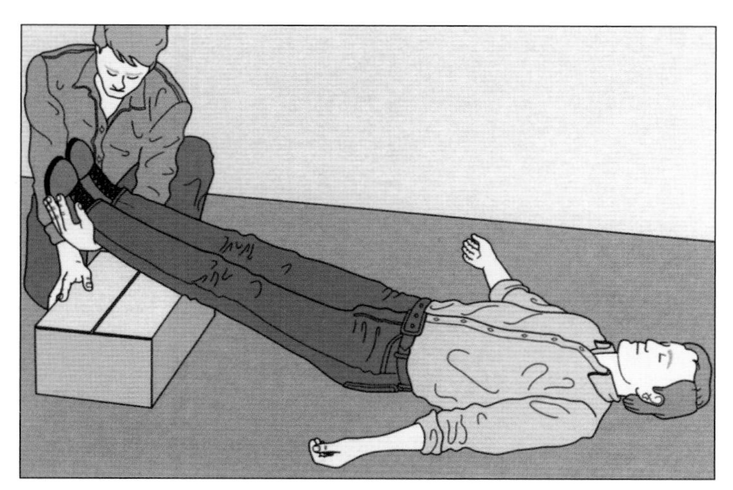

Handling a Shock Victim

Treating a shock victim

Some degree of shock—either immediate or delayed—accompanies any injury. Shock can be provoked by loss of blood, pain, or an allergic reaction. Signs of shock include anxiety or confusion; cold or clammy skin; weak, irregular breathing or pulse; and loss of consciousness. If you suspect an injury victim is suffering from shock, immediately call for emergency help. If the victim is conscious, place him on his back with his feet propped up 8 to 12 inches above the level of his head *(left)*. Loosen the victim's clothing around the neck, chest, and waist. Keep the victim warm with a blanket, but avoid overheating. Do not give the victim anything to eat or drink.

Index

Index

Back to Basics

THE *Missing* SHOP MANUAL SERIES

These are the manuals that should have come with your new woodworking tools. In addition to explaining the basics of safety and set-up, each *Missing Shop Manual* covers everything your new tool was designed to do on its own and with the help of jigs & fixtures. No fluff, just straight tool information at your fingertips.

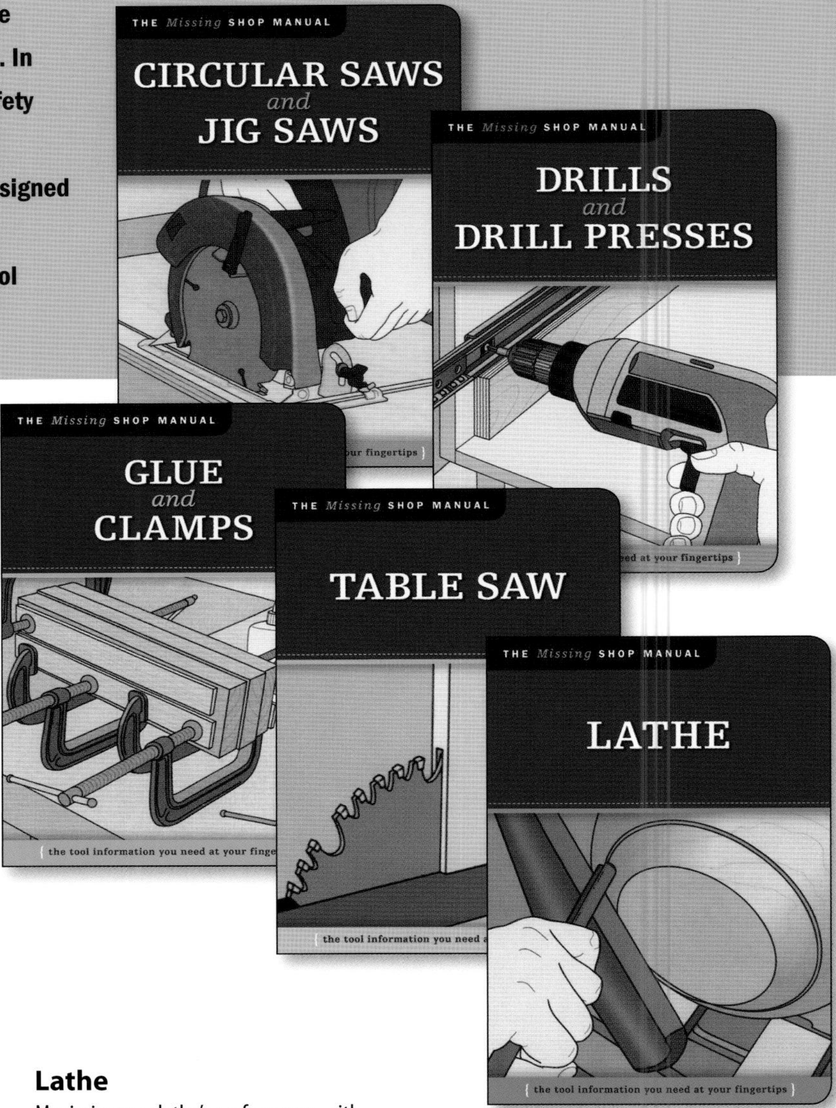

Circular Saws and Jig Saws

From ripping wood to circle cutting, you'll discover the techniques to maximize your saw's performance.

ISBN 978-1-56523-469-7
$9.95 USD • 88 Pages

Drills and Drill Presses

Exert tips and techniques on everything from drilling basic holes and driving screws to joinery and mortising.

ISBN 978-1-56523-472-7
$9.95 USD • 104 Pages

Glue and Clamps

Learn how to get the most out of your clamps and that bottle of glue when you're carving, drilling, and building furniture.

ISBN 978-1-56523-468-0
$9.95 USD • 104 Pages

Table Saw

Whether you're using a bench top, contractor or cabinet saw, get tips on everything from cutting dados and molding to creating jigs.

ISBN 978-1-56523-471-0
$12.95 USD • 144 Pages

Lathe

Maximize your lathe's performance with techniques for everything from sharpening your tools to faceplate, bowl, and spindle turning.

ISBN 978-1-56523-470-3
$12.95 USD • 152 Pages

Back to **Basics** *Straight Talk for Today's* **Woodworker**

Get *Back to Basics* with the core information you need to succeed. This new series offers a clear road map of fundamental woodworking knowledge on sixteen essential topics. It explains what's important to know now and what can be left for later. Best of all, it's presented in the plain-spoken language you'd hear from a trusted friend or relative. The world's already complicated—your woodworking information shouldn't be.

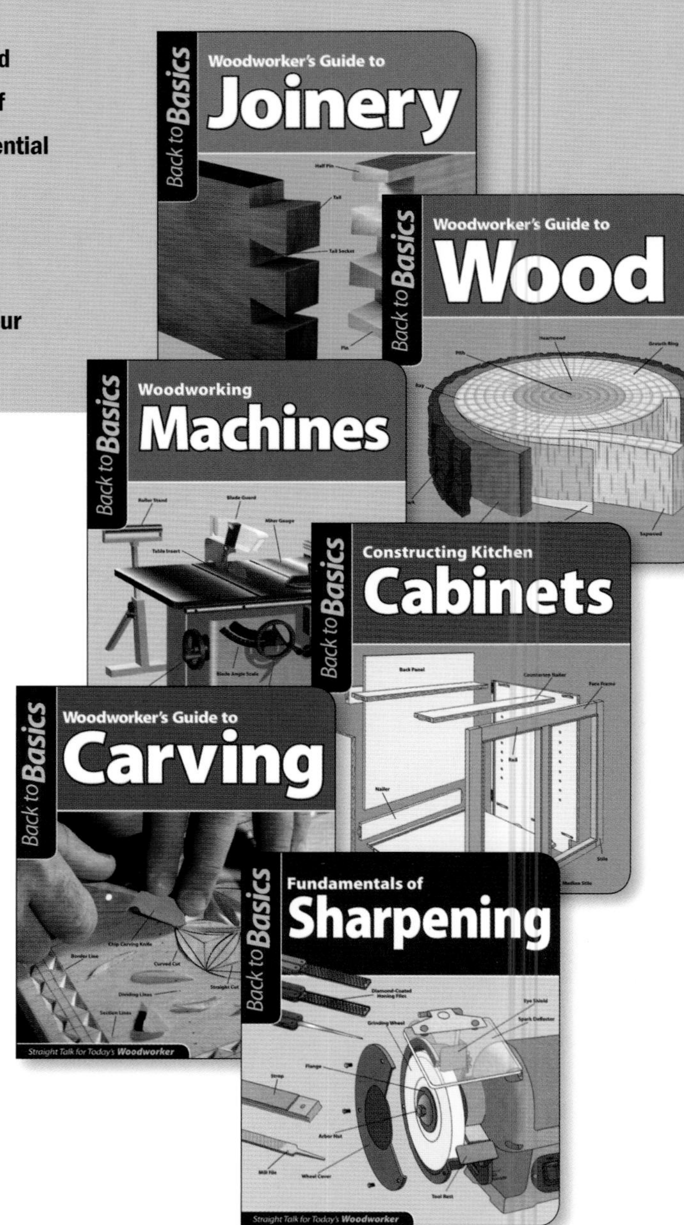

Woodworker's Guide to Joinery

ISBN 978-1-56523-462-8
$19.95 USD • 200 Pages

Woodworker's Guide to Wood

ISBN 978-1-56523-464-2
$19.95 USD • 160 Pages

Woodworking Machines

ISBN 978-1-56523-465-9
$19.95 USD • 192 Pages

Constructing Kitchen Cabinets

ISBN 978-1-56523-466-6
$19.95 USD • 144 Pages

Woodworker's Guide to Carving

ISBN 978-1-56523-497-0
$19.95 USD • 160 Pages

Fundamentals of Sharpening

ISBN 978-1-56523-496-3
$19.95 USD • 128 Pages

Look for These Books at Your Local Bookstore or Woodworking Retailer

To order direct, call **800-457-9112** or visit *www.FoxChapelPublishing.com*

By mail, please send check or money order + $4.00 per book for S&H to:
Fox Chapel Publishing, 1970 Broad Street, East Petersburg, PA 17520